To Simon

International Perspectives on Teaching and Learning for Mathematical Literacy

Mathematics Teaching and Learning

Series Editor

Yeping Li (*Texas A&M University, College Station, USA*)

VOLUME 6

The titles published in this series are listed at *brill.com/mtal*

International Perspectives on Teaching and Learning for Mathematical Literacy

Edited by

Oda Heidi Bolstad, Simon Goodchild and Merrilyn Goos

BRILL

LEIDEN | BOSTON

Cover illustration: iStock.com/Wannaludee Usanakul

All chapters in this book have undergone peer review.

The Library of Congress Cataloging-in-Publication Data is available online at https://catalog.loc.gov

Typeface for the Latin, Greek, and Cyrillic scripts: "Brill". See and download: brill.com/brill-typeface.

ISSN 2589-6016
ISBN 978-90-04-50450-9 (paperback)
ISBN 978-90-04-50451-6 (hardback)
ISBN 978-90-04-71096-2 (e-book)
DOI 10.1163/9789004710962

This book is printed on acid-free paper and produced in a sustainable manner.

Contents

List of Figures and Tables IX
Notes on Contributors XII

1 International Perspectives on Teaching and Learning for Mathematical Literacy: An Introduction 1
Oda Heidi Bolstad and Simon Goodchild

PART 1
Mathematical Literacy in Teaching and Teacher Education

2 Preparing Pre-service Teachers to Engage in Mathematical Literacy Practices across the Curriculum 23
Kathy O'Sullivan and Merrilyn Goos

3 The Teaching of Mathematical Literacy in Four South African Classrooms 46
Ellamma Pillay and Sarah Bansilal

PART 2
Designing and Evaluating Mathematical Literacy Tasks

4 Levels of Context Use: An Analytical Tool to Inform the Influence of Problem Context on Students' Performance and to Construct Mathematical Problems 71
Felipe Almuna and Kaye Stacey

5 Interplay between Contextual and Mathematical Literacy in Task Design: The Math-LIGHT Program as an Example 95
Roza Leikin, Masha Boriskovsky and Math-LIGHT team

PART 3
Mathematical Literacy in Real World Tasks and Experiences

6 Supporting Teachers to Develop Mathematical Literacy through Contemporary Financial Contexts 125
Carly Sawatzki, Jill Brown and Tim Powers

7 Mathematical Literacy in the Context of Statistics: A Theoretical Discussion and an Example from a Research Project 152
Jonas Bergman Ärlebäck, Peter Frejd, Robert Schaller and Patrick Bergman Skiba

8 Fostering Student's Mathematical Literacy through Mathematical Modelling Tasks 178
Peter Frejd, Jonas Bergman Ärlebäck and Pauline Vos

PART 4
Commentary

9 Mathematical Literacy as a Disruptive Innovation 207
Merrilyn Goos

Index 221

Figures and Tables

Figures

2.1 Numeracy model (Goos et al., 2014). 27
3.1 Task based on lines with different gradients. 54
3.2 Question 4.5 focusing on the rainwater runoff from the roof (Reproduced from DBE, 2012, p. 6). 55
3.3 Volume of imaginary prism. 56
3.4 Ratio representation of the teacher-learner problem. 58
3.5 Extract from the textbook read by Grace (Reproduced from DBE, 2015). 60
3.6 Calculations for Question 2.1. 61
4.1 The PISA modelling cycle. 74
4.2 Modified Rock Concert PISA problem. 75
4.3 Item M266Q01 Carpenter (OECD, 2006). 78
4.4 Item M465Q01 Water Tank (OECD, 2006). 79
4.5 Classifying problems according to levels of context use, by combining scores on two processes. 83
4.6 Two problems at first-level of context use (Almuna, 2020). 84
5.1 Hat problem. 96
5.2 Modelling cycle from educational modelling (based on Kaiser & Stender, 2013; the term "real life" is replaced by "context" by the author). 99
5.3 Findings from teachers' reports. 117
5.4 Solution to the Hat problem. 119
6.1 BNPL terms and conditions. 136
6.2 The teachers' disciplinary affiliations (n = 47). 137
6.3 Teachers' ratings on the usefulness of aspects of the course (n = 24). 139
6.4 Indicators of teacher learning and agency as a result of the course (n = 24). 140
7.1 A framework for ML, recreated based on Geiger et al. (2014). 154
7.2 The investigative cycle (dimension 1) in Wild and Pfannkuch's *framework for statistical thinking* (Wild & Pfannkuch, 1999, p. 226). 157
7.3 The interrogative cycle (dimension 3) in Wild and Pfannkuch's framework for statistical thinking (Wild & Pfannkuch, 1999, p. 226). 158
7.4 Arrows added to Figure 7.1 to illustrate how models (cf. Lesh & Doerr, 2003) fundamentally permeate and hold together the dimension and components of ML (building on the framework by Geiger et al. (2014)). 166
7.5 The slide the teacher used in discussing the notions of beat, pulse, and BPM. 168

8.1 Modelling cycle inspired by Blum (2015). 182
8.2 The six containers of beads representing the unknown populations (N = 20, 400, 10,000; two containers of each quantity – one uniformly distributed and one skewed). 188
8.3 Screen shots of (a) the default Map tab; (b) the Disease tab for evolving the disease; and (c) the World tab with a statistical summary and information about the progress of the cure being developed. 191
8.4 Sankey diagrams showing energy conversion in lightbulbs. 194
8.5 Sankey diagram designed by the students in the Project Waste Sorting. Text on the left: "Flow diagram from the school Wednesday 13th February 2019". Text in the middle: "13th February 2019: 867 unrecyclable waste: 867". Text on the right: "food waste: 382, glass and metal: 60, plastics: 253, unrecyclable waste: 91, carton and paper: 81". 196
9.1 The curriculum policy, design, and enactment system (Remillard & Heck, 2014, p. 709). 209
9.2 Numeracy model (Goos et al., 2014). 210
9.3 "N" Framework (O'Sullivan, 2022). 215

Tables

2.1 Examples of how numeracy can be embedded across the school curriculum. 29
2.2 Subject specialisms of pre-service teachers. 33
2.3 Phases of thematic analysis. 35
2.4 Cross tabulation of pre-service teachers' subject specialism against perception of adequacy of personal mathematical knowledge. 38
2.5 Information about courses offered by universities which incorporate numeracy. 39
3.1 Table of values that learners were asked to complete. 60
4.1 Distribution of ratings of 453 solutions from 151 students by levels of context use and statistics from ordinal logistic regression model (with zero-order level of context use as reference). 87
5.1 Analysis of target problem 1. 106
5.2 Analysis of target problem 2. 110
5.3 Analysis of target problem 3. 115
6.1 Teachers interviewed and their school community. 138
7.1 Types of thinking (dimension 2) in Wild and Pfannkuch's (1999) framework *for statistical thinking*: general types of thinking (G); and statistic-specific types of thinking (S). 157

7.2 The seven knowledge and disposition elements in Gal's (2002) framework of SL. 161
7.3 The SL construct by Watson and Callingham (2003) as summarized by Sharma (2017, p. 123). 162
7.4 The summary statistics for the class data as presented by the teachers on the smartboard. 170
8.1 The three modelling tasks. 186

Notes on Contributors

Felipe Almuna
Ph.D. (2020), The University of Melbourne, is Assistant Professor of Mathematics Education at Universidad Austral de Chile. He has published research articles on the influence of context of mathematical problems on the student's performance.

Jonas Bergman Ärlebäck
Ph.D. (2011), is Professor in Mathematics education in the Department of Mathematics at Linköping University, Sweden. His research interests are foremost aspects related to the teaching and learning of, and through, mathematical modeling, and the teaching and learning of statistics.

Sarah Bansilal
D.Maths Ed. (2008), is Professor of Mathematics Education at the University of KwaZulu-Natal. She has published numerous articles and book chapters which focus on teacher education in Africa, including "Exploring teachers' practices in teaching Mathematics and Statistics in KwaZulu-Natal schools" (South African Journal of Education, 2017).

Oda Heidi Bolstad
Ph.D. (2020), is Associate Professor of Mathematics Education at Volda University College, Norway. She is a teacher and teacher educator and her research areas include mathematical literacy and mathematical modelling.

Masha Boriskovsky
is a Ph.D. candidate at the Department of Mathematics Educations in the University of Hiafa. Her research focuses on the contribution of literacy tasks to the development of students' mathematical knowledge and problem solving skills as well as on the teachers' proficiency in teaching literacy tasks.

Jill Brown
designs courses that prepare preservice and practising teachers (including out-of-field teachers) to be research-informed mathematics educators. Her research in the field of mathematical modelling, including her learning task design, is internationally recognised.

Peter Frejd

Ph.D. (2014), Linköping University, is an associate professor of mathematics education in the Department of Mathematics. His research focuses on mathematical modelling, mathematical literacy and mathematics in the workplace.

Simon Goodchild

was Professor at the University of Agder, Norway. He was a mathematics teacher and teacher educator who published research on mathematics teaching in higher education. He was central in establishing MatRIC – Centre for Research, Innovation and Coordination of Mathematics Teaching, for which he also was the leader.

Merrilyn Goos

is Adjunct Professor of Education at the University of the Sunshine Coast, Australia. She is a mathematics educator who has published research and teacher education resources on numeracy across the curriculum and across the lifespan.

Roza Leikin

Ph.D. (1997), Technion – Israel Institute of Technology, is Full professor of mathematics education and education of gifted and talented students. She is the Dean of the Faculty of Education at the University of Haifa. She published over 180 peer reviewed research papers and 14 edited volumes forcusing on mathematical expertise, giftedness, and creativity as well as on teachers' and teacher educators' professional potential.

Kathy O'Sullivan

is a lecturer in Mathematics Education and Director for the Bachelor of Mathematics and Education programme at the University of Galway. She specialises in numeracy and mathematics education, initial teacher education and continuous professional development for teachers.

Ellamma Pillay

Ph.D. (2018), is an enthusiastic mathematics teacher at a school in South Durban. She has published a few articles, including "Negotiating the demands of teaching Mathematical Literacy" (Africa Education Review, 2019).

Tim Powers

leads Monash Statistical Consulting Services, mentoring researchers in statistical analyses. With a PhD in Educational Psychology and a Master of Applied

Statistics, Tim also supports university research in survey design, qualitative analysis, and project coordination.

Carly Sawatzki
supports teachers of mathematics to teach differently, by helping them to connect students' classroom learning with the real world. Carly is internationally recognised for her thought leadership on young people's financial capability [financial literacy/financial numeracy].

Robert Schaller
has been a mathematics and music teacher at the lower secondary level for 30 years. Presently he is working at Djäkneparkskolan in Norrköping, Sweden, and is frequently involved in developmental and research projects.

Patrick Bergman Skiba
has been a mathematics and english teacher at the lower secondary level for 10 years. Presently he is working at Navestadsskolan in Norrköping, Sweden, and is frequently involved in developmental and research projects.

Kaye Stacey
is Emeritus Professor of Mathematics Education at The University of Melbourne. Her research work is renowned for its high engagement with schools. Her doctoral thesis from the University of Oxford, UK, is in number theory. She was the Chair of the Mathematics Expert Group for the OECD's 2012 PISA survey. Kaye Stacey was awarded a Centenary Medal from the Australian government for outstanding services to mathematical education.

Pauline Vos
Ph.D. (2002), is Professor of Mathematics Education in the Department of Language, Literature, Mathematics and Interpreting at the Western Norway University of Applied Sciences. She leads practical research into the design of learning trajectories that deviate from deductive, teacher-centred approaches, and theoretical research into the why, when and how of using/learning what type of mathematics by which students/users.

CHAPTER 1

International Perspectives on Teaching and Learning for Mathematical Literacy

An Introduction

Oda Heidi Bolstad and Simon Goodchild

Abstract

This chapter sets out the rationale and purpose of the book. A preliminary explanation of the nuances of meaning of mathematical literacy (ML) is offered, followed by a brief summary of research on ML carried out by the first author in Norway, the relevance of this to the international context is explained. Four issues about teaching and learning for ML faced by practitioners and researchers are presented: conceptualisation, professional development, teacher education, and educational resources for ML. The remainder of the chapter is composed of brief introductions to each of the remaining chapters in the book.

Keywords

mathematical literacy – numeracy – mathematics education

1 Prelude

This volume is composed of chapters written by internationally recognised researchers and practitioners who focus on teaching and learning mathematical literacy (ML). Their work combines inquiry into current practices and sustained practical intervention aimed at improving students' experience of learning for ML. The collected works, emerging from several educational systems, offer a fresh perspective on ML that will be of value to practicing and pre-service teachers, and researchers.

We believe this volume will be of value on teacher education programmes, for both mathematics and general courses since in many places ML is expected to permeate the whole curriculum. The chapters that follow include practical examples that pre-service teachers might adapt to their own needs, and

 | DOI:10.1163/9789004710962_001

descriptions of research that could inspire student projects at bachelor's and master's levels. It is arguable that all school subjects can include opportunities to develop ML in situations authentic to the discipline. Research reported later in this book reveals that many (perhaps the majority of) practicing teachers have not received any formal education for teaching ML, thus there is much herein that is of value to experienced practitioners in all subjects. Furthermore, as the contributors to this book are experienced researchers, we hope it will be of value to mathematics education researchers with an interest in gaining deeper knowledge of ML as a research field.

An underlying purpose of the book is to support the development of rich understandings and practices of how an education for ML can be realised in practice. We hope teacher educators in pre-service and in-service teacher education programmes will be supported in promoting education for ML by the chapters in this volume. The chapters provide practice-based experiences and research informed suggestions about how to support the development of students' ML. Also, we hope, teacher educators and teachers will be inspired to design and implement tasks in their own teaching and thereby provide students with learning opportunities to develop competencies in ML.

In the preparation of this volume, we invited contributions from researchers who we believed would provide an internationally representative collection of ML focused research. To ensure a parallel focus on practice, these invited researchers were encouraged to co-author with practicing teachers or colleagues working very closely with schoolteachers. Our hope, therefore, is that readers will find the contents of value in research-based practice as well as practiced-informed research.

2 Mathematical Literacy – What and Why?

It is an indisputable truism that mathematics plays an important role in today's world, for example, in finance, health, environment, and decision making. On a regular basis, members of modern society engage in interpreting and using quantitative data, graphical representations, information about shape, space, time, and uncertainty, mathematical language and symbolic statements. Mathematics is used in daily life more than is often appreciated. The competence to deal with these mathematically related aspects of engaging in the community with independence and agency is referred to using a variety of expressions that are presented in later chapters in this book, but here, as in the title, we mainly use the term mathematical literacy (or ML) which has been popularised especially through the international PISA studies. That is, the Programme for International Student Assessment (PISA), carried out under the auspices of the OECD. In the 2021 PISA analytical framework, mathematical literacy is defined as:

> an individual's capacity to reason mathematically and to formulate, employ, and interpret mathematics to solve problems in a variety of real-world contexts. It includes concepts, procedures, facts and tools to describe, explain and predict phenomena. It assists individuals to know the role that mathematics plays in the world and to make the well-founded judgments and decisions needed by constructive, engaged and reflective 21st century citizens. (OECD, 2018, p. 7)

The PISA framework has had a great impact on the development of the participating countries' curricula and policy documents and educational discourse (Hopfenbeck et al., 2018; Santos & Centeno, 2023). Even though the notion of ML may not be stated explicitly, a wide range of aspects of ML is implicit in all school curricula. Also, there exists several different conceptualisations of the notion of ML, and countries have organised its inclusion within their national curricula in different ways. For example, in Norway, ML is considered a cross-curricular competence whereas in South Africa, ML has been assigned the status of an identified subject in the school curriculum (see Chapter 3 by Pillay & Bansilal, this volume).

Although many definitions and descriptions of ML (and related notions) exist in curricula and policy documents (many of these are presented in later chapters), they do not provide guidance for teachers in terms of how to design and implement ML tasks in their teaching. Also, there is little consistency in the way teacher educators support prospective teachers in developing their ML teaching competence (see Chapter 2 by O'Sullivan and Goos, this volume). Therefore, we hope our focus on the challenges in enhancing ML and how to deal with these challenges, will make this book a valuable contribution to the body of literature in this field.

In recent years, there has been a growing understanding that under-developed mathematical competences limit an individual's prospects in terms of career opportunities, social well-being, financial security, and political participation. For instance, the study by Levy et al. (2014, p. 107) revealed that individuals with low ML competence are less likely to understand health risks or comply with medication regimes. The research points to those with low ML competence and their underuse of screening for colorectal cancer, their greater difficulty managing chronic conditions, and reporting worse subjective health, in comparison with those who have higher ML competence.

Mathematical knowledge is also required in employment in many different ways, as demonstrated, for example, by the anthropologist Lave (1988) and studies to which she refers. Straesser (2015) claims that elementary arithmetic is a part of work-related mathematics in each and every vocational context. However, what practitioners do and how they do it may not always be predictable from considerations of general mathematical methods (Noss et al., 2000). Many studies have investigated the mathematical knowledge related to different vocational contexts. For instance, Saló i Nevado and Pehkonen (2018) investigated Finnish cabinetmakers' use of mathematics and problem solving. Moreira and Pardal (2012) examined masons' professional practices in Portugal, describing the daily practices of masons and how geometry and arithmetic emerge from the masons' work tasks. Saló i Nevado et al. (2011) studied how Spanish farmers make use of mathematics in daily life situations on a farm in terms of distributing space in a barn for feeding calves and the use of different objects as measuring tools. These studies show that people use mathematics in their work, and particular occupations have specific requirements and tasks related to different kinds of mathematical knowledge.

Following the outbreak of the COVID-19 pandemic in early 2020, citizens' need for ML became evident on a global scale. The information provided to the public about the spread of the pandemic, locally, nationally, and internationally, were expressed in mathematical terms. These terms consisted of various types of graphs, tables, and concepts referring to mortality rate, numbers of people hospitalised, and the speed at which the virus was being transmitted, and mathematical modelling as a basis for prediction. Aguilar and Castaneda (2021) found that the mathematical competences needed to interpret the official COVID-19 information in Mexico comprised mathematical communication, mathematical representation, mathematical symbols and formalism, mathematical modelling, and mathematical reasoning competency. Numerous articles regarding the mathematical demands related to the COVID-19 pandemic can be found, for example in a special issue of *Educational Studies in Mathematics* (Chan et al., 2021). Other examples of societal issues requiring mathematical competence can be found in a special issue of *Journal of Mathematical Behaviour* (Mamolo & Thanheiser, 2023) and a ZDM issue on "Connecting citizenship education and mathematics education" (Maass et al., 2023).

Mathematics in everyday life, work life, and societal life always has an extra-mathematical purpose. This purpose is to cope with the various situations one encounters, and the ML competency required is determined by these situations. To ensure that students develop the competence to cope with the routine quantitative aspects of life and to foster an equitable society, ensuring that everyone has a realistic opportunity to develop competence in ML is surely an important educational goal for all countries and communities.

3 Researching ML in Norway

The background for initiating the work on this book was the first editor's research into the implementation of ML curriculum policies in three Norwegian secondary classrooms (Bolstad, 2020b). As a background to this study Bolstad with colleagues conducted a literature review focusing on research on ML in schools (Haara et al., 2017). Bolstad's study aimed to gain knowledge about the nature of teaching and learning for ML in Norway. However, the preceding review of research revealed that the situation in Norway is not atypical internationally, and this present volume is a reaction to her work in these endeavours.

In Norway, there has been great emphasis on students' competence in the STEM subjects for more than 20 years. Poor Norwegian results on international educational assessments, such as PISA and TIMSS, is part of the reason for this emphasis. Since the Norwegian "PISA shock" in 2001, several educational policy strategies have been initiated to improve students' competences (Haugsbakk, 2013; Kongelf, 2019).

In 2006, a major curriculum reform (LK06) was implemented; the new curriculum contained a set of basic skills similar to the competences assessed in the PISA framework (Kjærnsli & Olsen, 2013). One of these basic skills is ML. There is an explicit requirement that involves using mathematical knowledge in various contexts and considered important for making decisions in everyday life and society (The Norwegian Directorate for Education and Training, 2012, p. 14). The basic skills are considered as cross-curricular skills. However, the different subjects play complementary roles in developing the required basic skills. Mathematics is considered as a core subject in the curriculum and carries the main responsibility for developing students' ML. In Norway, as in many other countries, students' ML competence is measured through national ML tests conducted in grades 5, 8, and 9.

From the emphasis on ML in the Norwegian curriculum, it might be expected that developing ML is a prominent part of teacher education and classroom teaching practice. The schools participating in the research by Bolstad (2020b) were located in a county in which students regularly performed above average on the national ML test. Therefore, it was anticipated that the research would provide insights into examples of best practice teaching. However, the research revealed that teaching and learning for ML is subject to several challenges.

There are conceptual challenges with the Norwegian notion related to ML. In Norwegian, this notion is "rekning som grunnleggande ferdigheit". The word "rekning" corresponds to "computation" or "arithmetic". It is challenging because it involves various interpretations, such as technical computations, understanding and meaning involved in computations, and practical computations in everyday life. The word "grunnleggande" corresponds to "basic" and creates associations to elementary, fundamental knowledge. Finally, "ferdigheiter", which corresponds to "skills", has historically been understood as technical and routine symbol treatment. Thus "rekning som grunnleggande ferdigheit" is directly translated as "computation as basic skill" and hides several nuances of meaning, and consequently, the Norwegian notion of numeracy involves several issues that can result in mixed understandings and goals.

Although the teachers participating in Bolstad's study acknowledge the importance of developing students' ML and express ideas of how this can be achieved (Bolstad, 2019), their operationalisation of ML in the observed lessons revealed little correspondence with the teachers' expressed meanings (Bolstad, 2020a). The challenges that the teachers face relate to their inadequate knowledge of how to enable students to experience an integrated *across the curriculum* opportunity to learn, and the embedded educational culture that entails heavy reliance on textbooks. Further, the students participating in the study appeared to approach mathematics as a subject that is primarily something done in school (Bolstad, 2023). Students' experiences of mathematics in everyday life involve specific topics and situations where the mathematical concepts and procedures are explicit and visible. Hence, as the teaching does not emphasise the uses of mathematics in the world unseen by the students, the role mathematics plays in education for citizenship is not emphasised.

As noted above, this volume is an outcome of Bolstad's experience of researching provision for ML in Norway, and her engagement with the international literature indicating that experiences in Norway are not dissimilar to those in other countries. It is our belief that there would be considerable value in creating an opportunity to share examples of international research and practice. In general, the meaning of elaborations on ML opportunities in the curriculum may appear clear for mathematics teachers, although we acknowledge there may be some variations in those meanings. However, it is almost certain that many teachers of other (non-numerate) subjects may find ML difficult to interpret in their own curriculum contexts. Further, if teachers rely on such ML opportunities identified in curriculum documents, they are unlikely to see additional ML learning opportunities that exist in every subject. It is to a large extent up to the teacher to identify opportunities and to develop learning activities related to ML. We hope this book may support teachers in the challenge of identifying these opportunities.

4 Mathematical Literacy Research and Practice in 2023: A Position Statement

Research on ML has shown that there are several challenges that need to be addressed. We briefly address a number of key issues as an overture to the following chapters in this volume.

4.1 *Conceptualisation*

Yasukawa et al. (2018) claim that there has been progress in terms of how ML is conceptualised with respect to historical interpretations of the concept. However, we are aware that there is still confusion concerning the character and nature of ML as a desirable attribute of every member of society. For example, Carter et al. (2015) identified five possible interpretations of ML exposed in schools in the State of Queensland, Australia in addition to the definition of "numeracy" set out in the Australian curriculum. First, ML is associated with competence in basic arithmetic. This is part of the problem in Norway, as previously described, and may have a historical explanation.

Second, in some schools, ML (or numeracy) is the name given to a curriculum programme in remedial mathematics. This may suggest to teachers of other subjects that pupils are already learning ML, and, hence, they do not need to include it in their own subject areas.

Third, there is an increasing practice of labelling primary and lower secondary school mathematics courses as "numeracy". For example, in South Africa, "mathematical literacy" is a compulsory subject for pupils who are not studying mathematics in grades 10–12, aiming to help them develop competence to understand and engage with mathematics in the world beyond the school boundary. If the school subject is associated with the conceptualisation of ML as basic arithmetic skills, school subjects with very different content will make a similar association. This has serious consequences for the understanding of ML.

Fourth, the production of curriculum support materials for teaching ML that closely resemble mathematics teaching resources can cause some teachers to regard "mathematics" and "mathematical literacy" as synonymous. The case of teaching resources is discussed later in this chapter.

Finally, referring to assessments as, for example, a "mathematical literacy test" (such as the NAPLAN test in Australia and the ML test in Norway) encourages teachers to believe that these tests exemplify the definition of ML. According to Carter et al. (2015), these constructions blur the distinction between school mathematics courses and ML. Not having a common interpretation of ML within a school, will have implications for the discourses and practices.

4.2 *Professional Development*

Professional development programmes can have a positive influence on teachers' understanding of ML and teaching. Goos and colleagues (i.e. Geiger et al., 2015; Goos et al., 2011, 2014) report results from professional development programmes in Australia. The professional development approach included three whole-day workshops, two rounds of school visits for lesson observations, teacher and student interviews, and the collection of student work. The teachers also completed written surveys regarding their confidence in ML teaching. The curriculum subjects for which the participating teachers were responsible were not restricted to mathematics. Participation in the project increased the teachers' confidence in terms of recognising the ML learning opportunities and demands in their own curriculum areas. Their confidence also increased in terms of determining students' ML learning needs to inform planning, demonstrating effective ML teaching strategies, and modelling ways of dealing with the ML demands of their curriculum area (Goos et al., 2014).

In a study by Liljedahl (2015), the participating mathematics teachers were brought together to develop a set of tasks for use within ML assessments. The teachers were engaged in discussions, sharing of experiences, designing, testing, and revising tasks. Although this was not intended as a professional development programme, the work led the teachers to make changes in their own teaching practice.

Changing practice is challenging, and the social interaction involved in the studies mentioned above, both in terms of working with other teachers and working with their students, appears to be a significant factor in the observed change process. It may be difficult to find the time and energy to engage in such collegial discussions, and, therefore, the challenges involved in developing ML tasks may seem too big to handle on top of all other issues the teachers have to deal with. However, according to Liljedahl (2015), designing ML tasks makes a crucial contribution to developing an understanding of the nature and scope of mathematical literacy.

4.3 *Teacher Education*

Another important issue that needs to be addressed is how prospective teachers are prepared to teach for ML through their teacher education programme. Although some countries may be fairly prescriptive in specifying the criteria to be satisfied for the recognition of qualified teacher status, we believe it is fairly common practice that individual universities and university colleges, decide the content of teacher education programmes, and course lecturers determine how to deal with developing prospective teachers' ML teaching competence. As in compulsory school, the main responsibility for developing prospective teachers' ML teaching competence might be attributed to the *mathematics* teacher educator. This means that students studying to become teachers of other subjects do not develop their competence to teach ML across the curriculum. According to a Norwegian research report (Finne et al., 2017) prospective teachers have limited knowledge and competence in how to implement the basic skills across subjects and in the different school grades. These results, in light of the rather general statements in the national teacher education guidelines, may suggest that more attention is needed in teacher education in order to prepare prospective teachers for teaching the basic skills generally, and ML specifically. This issue is explored in greater depth in the chapter by O'Sullivan and Goos (Chapter 2 this volume).

There are examples of teacher education programmes involving compulsory ML courses. In the Australian context, Forgasz et al. (2017) describe the Numeracy for Learners and Teachers course. It is anticipated that this will support prospective teachers in becoming practising teachers who consciously incorporate ML in their teaching.

The situation in South Africa is quite different. Here, ML is a subject distinct from mathematics. Teachers teaching ML in South Africa have not received specific training to teach the subject. Instead, it is mathematics teachers and teachers of other subjects who teach ML. More detail about the approach taken in South Africa is provided in the chapter by Pillay and Bansilal (Chapter 3 this volume). The varying emphasis on ML in teacher education programmes around the world also contributes to varying emphasis on teaching and learning for ML in schools.

4.4 *Educational Resources*

The curriculum is the most important guideline for what teachers are required to teach. In terms of providing support in teaching for ML, the Norwegian curriculum is limited. The curriculum provides a definition of Mathematical Literacy, although as stated earlier, as presented in the curriculum the notion is open to several nuances of meaning. This definition focuses on what students can do if they have this competence, and not what teachers can do to help them develop it, for example in terms of tasks or methods. An analysis of 12 national curricula (Frejd & Geiger, 2017) show that the idea of ML is represented in a limited fashion. This makes implementation difficult for the teacher.

Some online versions of national curricula provide support in teaching for ML in terms of filters connecting ML to subject topics. For example, in the online version of the Australian curriculum, the ML demands are explicitly identified by icons and online filters but the ML learning opportunities are not visible unless one knows what to look for (Goos et al., 2012). Hence, it is up to the teacher to identify relevant contexts and find suitable tasks to use in the classroom, and how the teacher deals with these challenges affects the opportunities students have to learn ML. In this way, the teacher is, to a large extent, on their own when it comes to identifying ML opportunities in the curriculum. However, if teachers rely only on ML opportunities identified in curriculum documents, they are unlikely to see additional ML learning opportunities that exist in all subjects. Furthermore, although the meaning of elaborations on ML opportunities in the curriculum may be clear for mathematics teachers, teachers of other subjects may find them difficult to interpret in their own curriculum contexts.

The lack of guidance in the curriculum regarding how to teach for ML contributes to the dominant use of the textbook. Historically, the textbook has been central in the development of teaching and learning practices in school, mediating between the intended and enacted curriculums. Teachers rely on textbooks in their teaching, even though they do not perceive this to be the best way to teach for ML. They feel that they lack the competence to teach ML across subjects (Bolstad, 2019), and they also expect that textbook authors have ensured that ML is embedded in the textbooks (Carter et al., 2015).

In addition, the increasing access to different digital resources has led teaching to become a hybrid practice where several resources work together, and teachers are more often designing their own tasks and activities to meet their students' needs (Rasmussen & Lund, 2015). Technological development has also led to several new ways of supporting teachers. Digital textbooks and textbook supplements, interactive boards, various software, and internet pages provide opportunities for educators to share their activities and experiences.

Although resources supporting teaching for ML exist, in fairly explicit ways, the teachers still need to rely on their own competence in order to ensure that they teach for ML. They have to evaluate the resources they choose to use and decide whether they are suited to serve the learning goals. Extensive use of, for example, textbooks confine the understanding of what ML is and how it can be developed because the development of such understanding is supported by designing, implementing, and revising ML tasks. By relying on teaching resources, teachers are transmitting someone else's understanding of ML. Also, inclusion of ML content in a textbook or a work programme does not ensure that students are given the opportunity to learn that content or develop ML competence.

Teachers may find it difficult to come up with their own ideas for suitable ML tasks. Geiger (2018) identified that an in-depth knowledge of the curriculum across subjects helped one teacher generate ideas for teaching across subjects. In this approach, the teacher looked for ideas by making connections between curriculum goals of different learning areas. Hence, the curriculum was used as a lens to look for teaching ideas. The approach taken by the other teacher was to connect teaching ideas to the curriculum and park them until there was a suitable time to use them. In this approach, the curriculum was not used as a lens, but rather as a way of facilitating a broader educational purpose. The ideas were already identified and fitted to serve the requirements of the curriculum.

It is our hope that the several chapters in this book that are illustrated with ML tasks designed and implemented in regular classrooms will contribute to meeting teachers' need for examples of ML tasks and approaches across the curriculum.

4.5 *Summary*

Teacher education can aim to develop students' knowledge domain so that they acquire sufficient curricular content knowledge, pedagogical knowledge, and mathematical content knowledge to embed ML in the subjects they study. And this means that teacher educators from all subject areas need some mathematical competence and to demonstrate how it relates to their subjects. How can novice teachers be expected to teach for ML across subjects if their teacher educators do not emphasise this and model how it can be done? There is no simple answer to the question of how teachers can be prepared for and supported in teaching for ML. It may be tempting to propose a restructuring of teacher education and implementation of teacher professional development programmes. However, such measures usually require significant resources in terms of time and funding, at least if they are attempted at large scale.

Moreover, preparing and supporting teacher development is complex because it is influenced by several actors. Policy makers, textbook authors, teacher educators, school owners, school leaders, in-service providers and prospective teachers all have a responsibility to work together to prepare for and support each other in terms of becoming instrumental in embedding ML in the school curriculum experienced by students. An essential ingredient is teachers' own motivation to develop their ML teaching competence. Therefore, researchers and teacher educators have an important job in raising their own and others' awareness of factors that can influence teacher motivation to develop ML competence and practice, and to create interest and support so that preservice and in-service teachers can become further engaged in teaching for ML across subjects.

Successful embedding of ML across school subjects requires professional development of teachers so that they gain experience with designing, testing, and revising ML tasks and develop their understanding of mathematical literacy. This also implies a review of curriculum documents and pedagogical practices. For professional development to be successful, the commitment and support of school leaders and adequate funding to support these activities is also required (Carter et al., 2015). We further emphasise from the earlier presentation of ML in teacher education, there also needs to be explicit focus on teaching and learning for ML in pre-service teacher education courses. This focus cannot be restricted to mathematics teacher education. It has to be implemented across teaching subjects so that all prospective teachers can gain experience in seeing the ML demands and opportunities in their subject areas and plan teaching and learning activities accordingly.

Experiences from researching ML in Norway and the accompanying review of international literature reveals the challenges faced and that these are experienced globally. However, international research has shown examples of how some of these challenges can be met. To improve the teaching and learning for ML, it is important that knowledge is shared. Sharing knowledge and experience is the main purpose of this book. In the following, the content of the chapters included in the book are described. Our hope is that they will contribute to disseminating fresh insights about teaching and learning for ML internationally, and the inspiration to try and resolve the challenges experienced.

5 Introduction to the Remainder of This Volume

The intention of this book is to present the work of mathematics educators who have been actively researching ML, or an area that is explicitly argued to be either an element of ML or an approach to developing ML competence. These presentations are organised into three sections on the themes: ML in teaching and teacher education (2 chapters), designing and evaluating ML tasks (2 chapters), and ML in real world tasks and experiences (3 chapters). The book concludes with a summary chapter by Merrilyn Goos whose own research in ML has had a significant influence on the research of many contributors to this book.

5.1 *Section 1: Mathematical Literacy in Teaching and Teacher Education*

Mathematical literacy is grounded in everyday contexts experienced by children and adults in their daily lives. To prepare children for the mathematical demands of their everyday life, as it is currently experienced and later as adults, the curriculums of many countries include an explicit reference to the development of competence in mathematical literacy (or synonymous word or phrase). However, there is little consistency in the characterization of mathematical literacy across curriculums either in content or manner of inclusion.

The chapter by Kathy O'Sullivan and Merrilyn Goos begins by outlining the development of mathematical literacy in school curricula in Norway, UK, Canada, Australia, and Ireland. Although well dispersed around the globe, the outcomes of their mathematics curricula show a high degree of similarity. For example, Olsen's (2006) analysis of the PISA 2003 study revealed students' performances in these countries to be quite closely correlated, leading Olsen to refer to them, on the basis of languages used, as members of a "Nordic-English cluster" (Olsen, 2006, p. 38). The curricula reviewed set out requirements for the inclusion of mathematical literacy in students' education; nevertheless, O'Sullivan and Goos point to the lack of formal guidance about how the requirements might be implemented. They proceed to explain how the absence of guidance from policy makers has been addressed by mathematics education researchers especially in the Australian context. However, by contrast, they observe, "Even though numeracy is identified as a cross-cutting competence in the junior secondary curriculum in Ireland, little research has been carried out to explore how teachers are embedding numeracy in the subjects they teach" (p. 28). One small scale action research project from Ireland that explored how mathematical literacy might be embedded in a range of school subjects is briefly presented. The chapter moves on to consider how students in Irish universities are prepared for their future role through preservice teacher education. O'Sullivan and Goos report some findings from a survey that included three universities' post-graduate teacher education programmes, they note considerable differences across the programmes regarding the focus and attention given to mathematics literacy across the curriculum, and unsurprisingly considerable differences in the meanings about the theme expressed by the prospective teachers.

Ellamma Pillay and Sarah Bansilal report from a "top down" approach in South Africa, where mathematical literacy has been introduced as a separate subject in the curriculum. The chapter reports research that focuses on four teachers, two mathematics specialists and two teachers who are not mathematics specialists. In all four cases, for different reasons set out in the chapter, the four teachers did not implement the mathematical literacy curriculum as intended. Pillay and Bansilal assert that more effort to prepare teachers for this type of curriculum development is required. Further, although not stated explicitly, the chapter raises questions about the nature and content of professional development courses that could prepare teachers for this work. What is the specific nature of pedagogical content knowledge for mathematical literacy that integrates context and mathematics? What specialized content knowledge of the contexts and mathematics is required? If it is "specialized" would the content be better introduced in a course focusing on that specialism – by specialist teachers?

5.2 *Section 2: Designing and Evaluating Mathematical Literacy Tasks*

ML is about being able to use mathematics as a cultural tool in regular activities of everyday life beyond the classroom; thus useful mathematics will be employed within a broader experiential context. Within the classroom, contexts are introduced for a variety of reasons, such as to stimulate interest, motivate engagement, and give a sense of how mathematics might be applied. In the development of mathematical literacy, context is a fundamental aspect of the learning experience and intended outcome. However, the introduction of context does not make a task an authentic experience of using mathematics outside the classroom. At best it can only offer a pseudo-authenticity because it would be very difficult for a classroom task to incorporate the variety of context markers and the desire to know the answer that would be present in situations encountered in regular activity outside the classroom.

The chapter by Felipe Almuna and Kaye Stacey reports research that shows how the introduction of context contributes to the difficulty of classroom tasks, and consequently has a negative effect on students' performance. Their starting point is to examine critically the tasks set in the OECD PISA studies, and they describe their classification of context use in mathematics tasks. Their approach considers the role of context in the translation from the text of the task description into a mathematical model/problem, and then the backwards translation from the "solved" mathematical problem to its interpretation in the real world (task) context. They then report from research that used this classification system in assessing how context contributed to the challenge of mathematical tasks in a sample of 150 students. Almuna and Stacey conclude:

> it is worth bearing in mind that problem designers, assessors and teachers face complex issues when writing problems embedded in real-world contexts. The contribution of this chapter has been to unpack these complex issues by providing a deeper understanding of level of context use, and the factors that contribute to it. (p. 91)

Roza Leikin and Masha Boriskovsky also consider the role of context in mathematical tasks, but they take a different perspective to that adopted by Almuna and Stacey, thus providing an interesting contrast. Leikin and Boriskovsky set mathematical modelling at the centre of their work and write about the roles of "contextual literacy" (relating to the origin of problems in the extra-mathematical (real) world) and "mathematical" or "conceptual" literacy that relates to the mathematical proficiency underpinning the analysis of mathematical models. The chapter reports from a professional development project:

> the Math-LIGHT program, a program that integrates contextual and conceptual literacy aimed at the development of literacy of both kinds: the tasks are aimed at developing students' mathematical competencies by acquainting them with various phenomena in science and life, and broadening their contextual literacy through the solving of mathematical problems. (p. 97)

The chapter presents and discusses several tasks that have been designed for teachers use with secondary students at various levels of mathematical development. The tasks are used to explain how the designs are "stepped" (i.e., provided with a ladder rather than a scaffold) to facilitate students' entry into both context and mathematics.

Leikin and Boriskovsky include some results from a survey of teachers' evaluations of the tasks. They report on how teachers experience the role of the tasks, how the tasks support learning and motivate interest. In contrast with Almuna and Stacey, the focus is not about evaluating the level of difficulty that context adds to a task; rather, Leikin and Boriskovsky are concerned to present tasks that have the potential to develop students' contextual and mathematical literacy in tandem.

5.3 *Section 3: Mathematical Literacy in Real World Tasks and Experiences*

In regular everyday life one rarely meets tasks that are immediately identifiable with the mathematics of the classroom, other than perhaps classroom tasks that use so called "real-world" contexts. The mathematically literate person will be competent to apply mathematical knowledge in a range of situations – financial/economic, health, politics, etc. In Chapter 6, Carly Sawatzki, Jill Brown and Tim Powers consider the requirement for schools in Australia to educate students for "financial capability" as a particular aspect of numeracy (the term for mathematical literacy used in the Australian curriculum). The Australian school curriculum describes the learning outcomes for financial capability in general terms and even points to the curriculum areas that have responsibility for the development of students' knowledge of the topic (Humanities and Social Sciences, and Mathematics). Nevertheless, schools and teachers are left to interpret the curriculum requirements and ensure those responsible for teaching have the competencies (content and pedagogic knowledge) to ensure the requirements are met. The authors of this chapter start from the realization that schools and teachers are neither adequately prepared nor supported for this role and they describe a design-research based professional development project for secondary teachers to address the needs.

In the chapter, examples of tasks designed to provide opportunities to learn financial capability are provided. These tasks are presented to participating teachers through an online professional development course, "*Economics + Maths = Financial Capability*". The chapter reports after two iterations of the course, using the analysis of a post-course evaluation survey (24 participants) and semi-structured interviews (14 participants). The analysis points to the error in assuming teachers are sufficiently knowledgeable about financial capability to be competent to teach the topic. Further, the chapter points to content and pedagogical content knowledge that should be included in professional development opportunities for teachers who are expected to teach financial capability.

In the chapter "Mathematical literacy in the context of statistics" Jonas Bergman Ärlebäck, Peter Frejd, Robert Schaller, and Patrick Bergman Skiba focus on statistical literacy and explain why this is not identical to, nor even a subset of mathematical literacy. Nevertheless, it is recognised that a knowledge of mathematics is an element of statistical literacy. However, the very fact that the chapter is included within a volume about mathematical literacy indicates the editors share the authors' perception of a strong relationship between these categories of literacy. The close relation between ML and statistical literacy is similar to that embedded in the school curricum as data handling and probability are included with the mathematics content section. The authors of this chapter set out the types of statistical knowledge, awareness and disposition that a statistically literate person needs in the present data "drenched" world. In fact, it is not difficult to infer that the statistically illiterate person lacks basic competence to be an independent agent in society. The chapter presents a "models and modelling" approach, which the authors explain as being appropriate for teaching and learning both mathematical and statistical literacy. After explaining this perspective, the approach is illustrated with an example useful for the development of both mathematical and statistical literacy taken from a grade 7 Swedish mathematics class.

Peter Frejd, Jonas Bergman Ärlebäck, and Pauline Vos demonstrate that mathematical modelling can be an effective approach to developing mathematical literacy. It is argued that learning mathematics out of context does not prepare a person for using it in context, and that mathematical modelling provides an opportunity for contextualizing the use of mathematics in authentic situations. The authors also explain and use a notion of *fostering* to explain how mathematical modelling promotes students' learning. In this chapter mathematical literacy is addressed from a perspective presented by Jablonka (2003), that is, from complementary aspects of the functionality of mathematical literacy regarding five different socio-cultural goals: development of human capital, maintenance of cultural identity, pursuit of social change, change in environmental awareness and evaluation of mathematical applications. The authors then present examples from their own research within mathematical modelling to illustrate how these five functional goals of mathematical literacy can be achieved at lower and upper secondary grades and within teacher education. The situations presented for mathematical modelling concern opinion polls of voting intentions, predictions of virus spread in a pandemic, and the use and interpretation of graphical displays "to visualise the flows of people, money, fluids, and so forth" (p. 194). In each case students are engaged actively in the situation modelled, either by practical or digital simulation or following a visit to an industrial enterprise.

In Chapter 9, Merrilyn Goos discusses ML as "disruptive innovation" in respect of curriculum design and enactment, and teacher preparation. Goos writes "'Disruption' is not meant to convey negative connotations, but instead to highlight the transformative possibilities of embracing mathematical literacy as a guiding principle in educating young people for an uncertain world" (p. 217). Her starting point is a curriculum model based on Remillard and Heck (2014), and a "Numeracy model" developed in the Australian context. The other chapters in this book are used to illustrate the issues upon which Goos focuses. The chapter closes with some questions that might be addressed in future research.

5.4 *Conclusion*

The chapters in this book share a common focus on approaches to teaching and learning for mathematical literacy. However, the chapters are also complementary in content and the approach taken to develop ML. They use various theoretical understandings of mathematical literacy, such as the 21C numeracy model (Geiger et al., 2015), the five socio-cultural goals of ML (Jablonka, 2003) and the PISA framework (OECD, 2018). The authors connect these understandings to other related educational concepts, such as mathematical modelling, statistical literacy, and contextual literacy. The settings in which the research have been conducted span from compulsory (middle) school, through secondary school, teacher education and teachers' professional development. Also, the tasks presented in the chapters provide insights into a variety of considerations and approaches one can take when working to embed mathematical literacy in the opportunities teachers create for students' learning. We hope, and believe, that the chapters will further provide ideas for how teaching and learning for ML can be enacted in classrooms and inspire teachers and teacher educators to implement these ideas.

References

Aguilar, M. S., & Castaneda, A. (2021). What mathematical competencies does a citizen need to interpret Mexico's official information about the COVID-19 pandemic? *Educational Studies in Mathematics, 108*(1), 227–248. https://doi.org/10.1007/s10649-021-10082-9

Bolstad, O. H. (2019). Teaching for mathematical literacy: School leaders' and teachers' rationales. *European Journal of Science and Mathematics Education, 7*(3), 93–108. https://doi.org/10.30935/scimath/9537

Bolstad, O. H. (2020a). Secondary teachers' operationalisation of mathematical literacy. *European Journal of Science and Mathematics Education, 7*(3), 115–135. https://doi.org/10.30935/scimath/9551

Bolstad, O. H. (2020b). *Teaching and learning for mathematical literacy* [Doctoral thesis, University of Agder]. Kristiansand. https://uia.brage.unit.no/uia-xmlui/handle/11250/2687234

Bolstad, O. H. (2023). Lower secondary students' encounters with mathematical literacy. *Mathematics Education Research Journal, 35*, 237–253. https://doi.org/10.1007/s13394-021-00386-7

Carter, M. G., Klenowski, V., & Chalmers, C. (2015). Challenges in embedding numeracy throughout the curriculum in three Queensland secondary schools. *The Australian Educational Researcher, 42*, 595–611. https://doi.org/10.1007/s13384-015-0188-x

Chan, M. C. E., Sabena, C., & Wagner, D. (Eds.). (2021). Mathematics education in a time of crisis – A viral pandemic [Special issue]. *Educational Studies in Mathematics, 108*(1–2), 1–13. https://doi.org/10.1007/s10649-021-10113-5

Finne, H., Mordal, S., & Ullern, E. F. (2017). *Oppfatninger av kvalitet i lærerutdanningene 2016* [Views on quality in teacher education 2016]. http://hdl.handle.net/11250/2569677

Forgasz, H., Leder, G., & Hall, J. (2017). Numeracy across the curriculum in Australian schools: Teacher education students' and practicing teachers' views and understandings of numeracy. *Numeracy, 10*(2), 1–20. https://doi.org/10.5038/1936-4660.10.2.2

Frejd, P., & Geiger, V. (2017). Exploring the notion of mathematical literacy in curricula documents. In G. A. Stillman, W. Blum, & G. Kaiser (Eds.), *Mathematical modelling and applications* (pp. 255–263). Springer International Publishing AG. https://doi.org/10.1007/978-3-319-62968-1_22

Geiger, V., Forgasz, H., & Goos, M. (2015). A critical orientation to numeracy across the curriculum. *ZDM – Mathematics Education, 47*(4), 611–624. https://doi.org/10.1007/s11858-014-0648-1

Goos, M., Dole, S., & Geiger, V. (2011). Improving numeracy education in rural schools: A professional development approach. *Mathematics Education Research Journal, 23*(2), 129–148. https://doi.org/10.1007/s13394-011-0008-1

Goos, M., Dole, S., & Geiger, V. (2012). Auditing the numeracy demands of the Australian curriculum. In J. Dindyal, L. P. Cheng, & S. F. Ng (Eds.), *Mathematics education: Expanding horizons. Proceedings of the 35th annual conference of the Mathematics Education Research Group of Australasia* (pp. 314–321). MERGA.

Goos, M., Geiger, V., & Dole, S. (2014). Transforming professional practice in numeracy teaching. In Y. Li, E. A. Silver, & S. Li (Eds.), *Transforming mathematics instruction: Multiple approaches and practices* (pp. 81–102). Springer International Publishing. https://doi.org/10.1007/978-3-319-04993-9_6

Haugsbakk, G. (2013). From Sputnik to PISA shock – New technology and educational reform in Norway and Sweden. *Education Inquiry, 4*(4), 607–628. https://doi.org/10.3402/edui.v4i4.23222

Hopfenbeck, T. N., Lenkeit, J., El Masri, Y., Cantrell, K., Ryan, J., & Baird, J.-A. (2018). Lessons learned from PISA: A systematic review of peer-reviewed articles on the programme for international student assessment. *Scandinavian Journal of Educational Research, 62*(3), 333–353. https://doi.org/10.1080/00313831.2016.1258726

Jablonka, E. (2003). Mathematical literacy. In A. J. Bishop, M. A. Clements, C. Keitel, J. Kilpatrick, & F. K. S. Leung (Eds.), *Second international handbook of mathematics education* (pp. 75–102). Kluwer Academic Publishers.

Kjærnsli, M., & Olsen, R. V. (Eds.). (2013). *Fortsatt en vei å gå: Norske elevers kompetanse i matematikk, naturfag og lesing i PISA 2012.* [*Still a way to go: Norwegian students' competence in mathematics, science, and reading in PISA 2012*]. Universitetsforlaget.

Kongelf, T. R. (2019). *Matematisk innhold og matematiske metoder i lærebøker brukt på ungdomstrinnet i Norge. Gullgruve eller fallgruve for utvikling av matematisk kompetanse i problemløsning og algebra?* [*Mathematical content and mathematical methods in textbooks used in lower secondary school in Norway*] [Doctoral dissertation, University of Agder]. Kristiansand.

Lave, J. (1988). *Cognition in practice: Mind, mathematics and culture in everyday life.* Cambridge University Press.

Levy, H., Ubel, P. A., Dillard, A. J., Weir, D. R., & Fagerlin, A. (2014). Health numeracy: The importance of domain in assessing numeracy. *Medical Decision Making, 34*(1), 107–115. https://doi.org/10.1177/0272989x13493144

Liljedahl, P. (2015). Numeracy task design: A case of changing mathematics teaching practice. *ZDM – Mathematics Education, 47*(4), 625–637. https://doi.org/10.1007/s11858-015-0703-6

Maass, K., Geiger, V., Gal, I., & Graven, M. (Eds.). (2023). Connecting citizenship education and mathematics education [Special issue]. *ZDM, 55*(4).

Mamolo, A., & Thanheiser, E. (Eds.). (2023). Mathematics in society: Exploring the mathematics that underpins social issues [Special issue]. *The Journal of Mathematical Behavior, 68.*

Moreira, D., & Pardal, E. (2012). Mathematics in mason's workplace. *Adults Learning Mathematics: An International Journal, 7*(1), 32–48.

Noss, R., Hoyles, C., & Pozzi, S. (2000). Working knowledge: Mathematics in use. In A. Bessot & J. Ridgway (Eds.), *Education for mathematics in the workplace* (pp. 17–35). Kluwer.

OECD. (2018). *PISA 2022 mathematics framework (draft)*. Retrieved October 20, 2021, from https://pisa2022-maths.oecd.org/

Olsen, R. V. (2006). A Nordic profile of mathematics achievement: Myth or reality? In J. Mejding & A. Roe (Eds.), *Northern lights on PISA 2003: A reflection from the Nordic countries* (pp. 33–45). Nordic Council of Ministers. https://www.regjeringen.no/globalassets/upload/kilde/kd/rap/2006/0006/ddd/pdfv/281852-pisa_northern.pdf

Rasmussen, I., & Lund, A. (2015). Læringsressurser og lærerrollen – Et partnerskap i endring? *Acta Didactica Norge, 9*(1). https://doi.org/10.5617/adno.2352

Remillard, J., & Heck, D. (2014). Conceptualizing the curriculum enactment process in mathematics education. *ZDM Mathematics Education, 46*, 705–718. https://doi.org/10.1007/s11858-014-0600-4

Saló i Nevado, L., Holm, G., & Pehkonen, L. (2011). Farmers do use mathematics: The case of animal feeding. *Nordic Studies in Mathematics Education, 16*(3), 101–121.

Saló i Nevado, L., & Pehkonen, L. (2018). Cabinetmakers' workplace mathematics and problem solving. *Vocations and Learning, 11*(3), 475–496. https://doi.org/10.1007/s12186-018-9200-8

Santos, Í., & Centeno, V. G. (2023). Inspirations from abroad: The impact of PISA on countries' choice of reference societies in education. *Compare: A Journal of Comparative and International Education, 53*(2), 269–286. https://doi.org/10.1080/03057925.2021.1906206

Straesser, R. (2015). "Numeracy at work": A discussion of terms and results from empirical studies. *ZDM – Mathematics Education, 47*(4), 665–674. https://doi.org/10.1007/s11858-015-0689-0

The Norwegian Directorate for Education and Training. (2012). *Framework for basic skills.* Ministry of Education and Research. https://www.udir.no/contentassets/fd2d6bfbf2364e1c98b73e030119bd38/framework_for_basic_skills.pdf

Yasukawa, K., Jackson, K., Kane, P., & Coben, D. (2018). Mapping the terrain of social practice perspectives of numeracy. In K. Yasukawa, A. Rogers, K. Jackson, & B. V. Street (Eds.), *Numeracy as social practice: Global and local perspectives* (pp. 3–17). Routledge.

PART 1

Mathematical Literacy in Teaching and Teacher Education

CHAPTER 2

Preparing Pre-service Teachers to Engage in Mathematical Literacy Practices across the Curriculum

Kathy O'Sullivan and Merrilyn Goos

Abstract

Mathematical literacy, referred to as numeracy in some countries, is a lifelong skill which students should develop in school. In some countries, such as Australia and Ireland, all teachers are required to embed mathematical literacy practices across the school curriculum. Despite this requirement for teachers to embed mathematical literacy learning across the school curriculum, there is little evidence of the best way to prepare teachers for this task. This chapter presents a review of policy developments and research in the field of mathematical literacy in teacher education. It also reports on a study conducted in Ireland, that explored the perceptions and experiences of pre-service secondary school teachers in relation to how well prepared they felt to embed mathematical literacy practices in different subjects across the school curriculum. The results showed that pre-service teachers need to be given the opportunity to learn about the concept of mathematical literacy and learn how to prepare to embed mathematical literacy practices in different subject disciplines.

Keywords

numeracy across the curriculum – mathematical literacy – pre-service teachers – numeracy policy

1 Introduction

Governments and policymakers around the globe have advocated that mathematical literacy is a lifelong skill which students should develop in school. As a result, mathematical literacy (sometimes known as numeracy or quantitative literacy) has been established as a cross-cutting competence within the school curriculum in many countries (Department of Education and Skills [DES],

 | DOI:10.1163/9789004710962_002

2011; Norwegian Directorate for Education and Training, 2012). Despite this requirement in schools for teachers to embed mathematical literacy across the school curriculum, there is little evidence of how pre-service teachers are prepared to embed mathematical literacy across the curriculum. This chapter aims to review policy developments and research in the field of mathematical literacy, and report on a study that explored the perceptions and experiences of pre-service secondary school teachers in relation to how well prepared they feel to embed mathematical literacy practices in subjects across the school curriculum in Ireland.

In many countries, the notion of mathematical literacy as a 21st century competency has emerged from international studies such as the OECD's Programme for International Student Assessment (PISA) (Niss, 2018). The PISA definition of mathematical literacy, initially conceived as the ability to use the mathematics learned in school and apply it to real world situations, has developed into a broader description as:

> an individual's capacity to reason mathematically and to formulate, employ, and interpret mathematics to solve problems in a variety of real-world contexts. It includes concepts, procedures, facts and tools to describe, explain and predict phenomena. It assists individuals to know the role that mathematics plays in the world and to make the well-founded judgments and decisions needed by constructive, engaged and reflective 21st century citizens. (OECD, 2018, p. 7)

In some English-speaking countries it is more common to speak of numeracy rather than mathematical literacy. The meaning of numeracy has also evolved over the years since the term first appeared in the 1950s in the Crowther Report, which investigated ways of improving the education of English boys and girls aged from 15 to 18 years (Ministry of Education, 1959). While the Crowther Report described numeracy simply as "the mirror image of literacy" (par. 398), numeracy is now understood to involve connecting "the mathematics learned at school with out of school situations that additionally require problem-solving, critical judgment, and making sense of the non-mathematical context" (Goos, 2020, p. 105).

Although mathematical literacy and numeracy have different historical roots, they can be regarded as closely related concepts that have influenced curriculum reform in many countries (Goos & O'Sullivan, 2023). For example, the school curricula in countries such as Canada, Australia, Ireland, and Norway identify mathematical literacy as a cross-curricular competence that should be developed in all subjects rather than only in mathematics (Alberta

Education, 2019; Australian Curriculum, Assessment and Reporting Authority [ACARA], n.d.; National Council for Curriculum and Assessment [NCCA], 2021; Norwegian Directorate for Education and Training, 2012). Despite this curriculum mandate, education authorities in these countries provide teachers with little guidance as to how to recognise the mathematical literacy demands of the subjects they teach. To date, only limited research has investigated how to prepare and support teachers for their role facilitating mathematical literacy learning across the curriculum. Most of the available studies involving practising teachers in primary and secondary schools (e.g., Bolstad, 2020; Goos et al., 2014; Liljedahl, 2015). Much less is known about how initial teacher education can help future teachers understand and implement strategies for developing students' mathematical literacy in their subject specialisms. The aim of this chapter is to stimulate informed and productive thinking about how Initial Teacher Education (ITE) programs can support pre-service teachers in developing their knowledge and understanding of mathematical literacy across the school curriculum, so that they are prepared to embed these practices in the classroom.

2 Mathematical Literacy in the School Curriculum

More than twenty years ago, the Quantitative Literacy Design Team was formed in the US to look into the meaning of numeracy in contemporary society. (Note that, in this context, the terms quantitative literacy, numeracy, and mathematical literacy tend to be used interchangeably.) This group of scholars led by Lynn Arthur Steen argued that numeracy development for students must be situated in an interdisciplinary educational context, permeating the whole school curriculum (Steen, 2001). This goal has been taken up by curriculum authorities in Australia and Ireland, more so than in other countries. In Australia, the National Numeracy Review Report commissioned by the Council of Australia Governments (2008) recommended:

> That all systems and schools recognise that, while mathematics can be taught in the context of mathematics lessons, the development of numeracy requires experience in the use of mathematics beyond the mathematics classroom, and hence requires an across the curriculum commitment. (p. 7)

Over the next ten years following publication of the National Numeracy Review Report, a new Australian curriculum was developed with cross-cutting "general capabilities" that intersect with learning areas representing disciplinary knowledge, skills, and understanding (see ACARA, n.d.). Numeracy is one of these general capabilities. However, Goos (2018) argues that because the official curriculum offers inadequate advice on how numeracy can be developed in each learning area, there must be doubts as to whether its cross-curricular aspirations for numeracy will be fulfilled as intended.

A similar situation exists In Ireland, where poor performance in the Third International Mathematics and Science Study (TIMSS) (Beaton et al., 1996) and PISA mathematical literacy in 2009 (Shiel et al., 2016) led to the launch of a national strategy to improve literacy and numeracy amongst children and young people (Department of Education and Skills [DES], 2011). A revised curriculum framework for the junior secondary years soon followed, introducing a set of eight "key skills" that act as cross-cutting competencies that teachers are meant to embed in the learning outcomes of every subject. Numeracy is identified as one of these key skills (NCCA, 2021). Yet there is no explanation in any of the newly developed subject specifications as to how numeracy or the other key skills can be recognised and developed in specific curriculum contexts.

3 Supporting Teachers to Embed Mathematical Literacy across the Curriculum

Despite curriculum reform processes endorsing the expectation that all teachers will be responsible for developing their students' mathematical literacy, it is often left up to researchers and professional development providers to offer "unofficial" guidance to teachers on how to address these curricular aims. Some of this research and development activity has yielded publications demonstrating how practising teachers were helped to recognise the mathematical literacy demands and opportunities of the subjects they teach. Most of this research has been carried out in Australia. For example, Thornton and Hogan (2004) worked with primary school teachers and secondary school non-mathematics teachers to set up classroom-based action research projects in which the teachers identified numeracy opportunities within lessons. This group of teachers met regularly with the researchers to share their observations of numeracy within all curriculum areas. From analysis of these discussions, Thornton and Hogan identified three idealised teacher orientations towards numeracy that they labelled separatist, theme-maker, and embedder. The separatist orientation described teachers who recognised the importance of mathematical skills but believed in leaving the teaching of these skills to the mathematics teacher. The theme-maker orientation was evident in teachers who developed schemes of work that integrated mathematics with other learning areas they thought would be relevant to the real world and, therefore, interesting to students. However, this approach, although common in the primary school years, risks trivialising one or more of the component learning areas and thus may undermine the integrity of the integrating theme. The embedder orientation, on the other hand, was found in teachers who could recognise the distinctive quantitative demands in all learning areas, were willing to engage in mathematical thinking that enhanced learning within their own subject, and believed that every teacher could be a teacher of numeracy without necessarily being an expert in mathematics. The three orientations identified by Thornton and Hogan hold some promise for working with other groups of teachers to help them recognise and refine their preferred approaches to embedding numeracy into the subjects they teach.

In another more extensive Australian research and development program, Goos and colleagues developed and tested a model of numeracy designed to capture relationships between mathematics, numeracy, other subjects taught in school, and real-world contexts (Goos et al., 2012). The model consists of four core elements: attention to real-life *contexts*; application of *mathematical knowledge*; use of physical, representational and digital *tools*; and promotion of positive *dispositions* towards the use of mathematics to solve real-world problems. A fifth overarching element – a *critical orientation* – requires the appropriate selection and application of mathematics to a real-world problem as well as the interpretation and critique of results (see Figure 2.1). These elements of the model together portray a rich conceptualisation of numeracy in the real world.

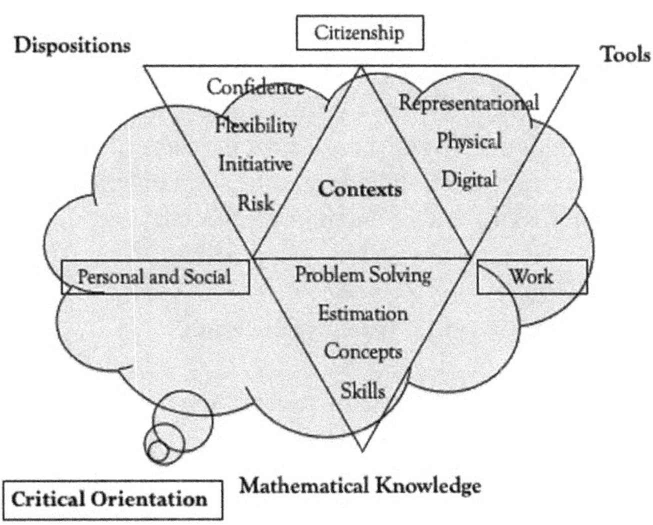

FIGURE 2.1 Numeracy model (Goos et al., 2014)

The numeracy model was designed to be accessible to teachers to support their planning and reflection. The usefulness of this numeracy model has been confirmed through several research and development projects involving Australian and Irish teachers in primary and secondary schools and across many subject domains. It has been effective in providing a framework for tracing teachers' growing understanding of numeracy across the curriculum, auditing the numeracy demands of the school curriculum, and helping teachers to recognise the numeracy demands and opportunities in different school subjects (Goos, 2021; Goos et al., 2012, 2014). Some examples of how teachers who participated in these projects embedded numeracy in different subjects are provided in Table 2.1 (for further details, see Goos et al., 2019).

Even though numeracy is identified as a cross-cutting competence in the junior secondary curriculum in Ireland, little research has been carried out to explore how teachers are embedding numeracy in the subjects they teach. In one recent example, Connolly et al. (2021) aimed to foster teacher collaboration and "explore how numeracy and mathematics can be integrated meaningfully and constructively in a range of carrier subjects" (p. 2). The term "carrier subject" highlights the inherent numeracy demands of subjects other than mathematics in the school curriculum. Carrier subjects provided contexts for making mathematical concepts more visible and meaningful to students. Using an action research methodology, the project team worked with pairs of secondary teachers in each participating school, where in each pairing one participant taught mathematics and the other taught a different discipline. Teachers designed numeracy investigations with the aim of achieving some overlap between the learning outcomes of the carrier subject and the mathematics curriculum. Although the project was initiated in the context of the national literacy and numeracy strategy (DES, 2011), this orientation to implementing numeracy practices aligns more closely with the traditions of interdisciplinary inquiry or curriculum integration than with approaches that take advantage of the inherent numeracy demands in disciplines other than mathematics (see Geiger et al., 2015, for a discussion of these different approaches).

4 Mathematical Literacy in Initial Teacher Education

When the school curriculum includes mathematical literacy as a cross-curricular competence, as is the case in Australia and Ireland, initial teacher education policies and standards likewise require that programs address pre-service teachers' personal mathematical literacy competence and their ability to promote mathematical literacy in the subjects they teach. The Australian

TABLE 2.1 Examples of how numeracy can be embedded across the school curriculum

Curriculum areas	Description of tasks	Curriculum goals
Health and physical education	Students wear a pedometer for a week and record their daily number of steps, converting steps to kilometres and representing the class results using graphs to look for and compare patterns of physical activity.	To monitor physical activity in order to design a personal fitness plan
English	Students read a selection of newspaper articles presenting statistical information in tables and graphs about the prevalence of dog attacks and plans for legislation imposing a ten-year ban from owning a dog on owners of dogs that attack people. They draw on these articles to write an opinion piece on the proposed new law.	To plan, draft and publish persuasive texts, selecting subject matter and language to convey information and ideas
History	Students construct a timeline for events that occurred during the Spanish conquest of the Americas, using an appropriate scale to highlight the sequencing, duration, and concurrency of events.	To use chronological sequencing to demonstrate the relationships between events and developments in different periods and places
Mathematics	Students evaluate two options for receiving pocket money: being paid either €45 per month or on a sliding scale of 10c on the first day of the month and increasing by 10c each day until the last day of the month. They justify the best scenario for this month, every month, and over the course of a year.	To recognise and solve problems involving rates and ratios

Institute for Teaching and School Leadership (AITSL), which plays a key role in regulating the teaching profession, has developed a professional standards framework that specifies what teachers at all career stages need to know and be able to do. The graduate standard relating to teachers' numeracy capabilities states that graduate teachers should "know and understand literacy and numeracy teaching strategies and their application in teaching areas" (AITSL, 2011, p. 13). Similarly, in Ireland, teacher education graduates must demonstrate knowledge and understanding of numeracy as it relates to school curriculum requirements (The Teaching Council, 2020). In both countries, this relatively weak framing of requirements for graduate teacher competence in promoting mathematical literacy does little to encourage universities to prioritise developing teacher education students' knowledge of strategies for embedding mathematical literacy across the school curriculum. As a result, it is not surprising to see diverse approaches to program design incorporating mathematical literacy into the teacher education curriculum. For example, while some universities offer semester-long courses on numeracy, sometimes in combination with literacy, others provide only one or two lectures on numeracy as part of a general pedagogy course or instead expect methods courses in the different areas of the school curriculum to address subject-specific numeracy demands. Thus, pre-service teachers in these two countries experience great variation in opportunities to learn mathematical literacy pedagogy. "Opportunity to learn" is a key indicator used in international comparative studies such as PISA and TIMSS to show whether students have access to high quality education. Here, "opportunity to learn" is used in the same sense as in the Teacher Education and Development Study in Mathematics (TEDS-M), a large-scale, international comparative study of the preparation of mathematics teachers, to refer to whether pre-service teachers have studied a particular topic in their teacher education program (Tatto et al., 2012).

The analysis of policy initiatives outlined above points to a lack of serious attention being given to mathematical literacy in initial teacher education in countries where mathematical literacy is, nevertheless, enshrined as cross-curricular responsibility for all teachers. Even so, accounts and evaluations of some exemplar Australian initiatives do exist. In an early example, Watson and Moritz (2002) explained how they created a Quantitative Literacy component as part of the mathematics course within a postgraduate teacher education program at their university. This component was based on a website they had previously developed, which compiled more than 200 newspaper articles containing significant mathematical content, together with teaching notes and student questions to support discussion of aspects of quantitative literacy in their daily lives (see Tasmanian Department of Education, 2019, for the current version of this website). Forty pre-service teachers completing primary, middle years, and secondary specialisations completed four assessment tasks requiring them to plan and teach lessons based on the website materials. While the quality of pre-service teacher responses varied, Watson and Moritz considered that the tasks were effective in introducing quantitative literacy into the teacher education curriculum.

Another early initiative, described by Groves (2001), involved developing a semester-long course in response to local teacher education guidelines before national professional standards were introduced in Australia. The course, titled *Numeracy across the curriculum*, targeted future secondary school teachers and addressed the nature and scope of numeracy in everyday life, the inherent demands and opportunities for numeracy in different subject specialisms, and teaching strategies to respond to school students' numeracy learning needs. The course was first offered in 1999 and enrolled 146 pre-service teachers who were preparing to specialise in 24 different secondary school subject areas. It was designed and taught by mathematics education academics at a time when there was little awareness, amongst teachers in schools and university staff teaching methods courses, of the role of numeracy within their subject areas. There were many obstacles encountered in delivering the course, such as the limited on-campus time available due to the scheduling school placement and pre-service teachers' apprehension about their personal numeracy abilities. Feedback from pre-service teachers was mixed, with some finding great value in the course and others clearly indicating their dislike for it. Perhaps the greatest challenge to engaging pre-service teachers was the lack of recognition, at that time, of the need for all secondary school teachers to take responsibility for development of students' numeracy capabilities.

A more recent example of implementing a mathematical literacy course for pre-service teachers is reported by Forgasz and Hall (2019). The impetus for designing the course arose from the inclusion of numeracy as a general capability in the Australian Curriculum and the introduction of numeracy standards for graduate teachers. The course, titled *Numeracy for learners and teachers*, was compulsory for all primary and secondary pre-service teachers. It provided an introduction to the meaning of numeracy, drawing on the model previously developed by Goos and colleagues to show pre-service teachers how to plan and implement numeracy-enriched tasks and lessons in science, financial literacy, persuasive writing, geography, history, the arts, technology, and health and physical education. The course was designed and coordinated by Forgasz, a mathematics educator, and taught in collaboration with teacher educators who specialised in other curriculum areas. It was evaluated for three consecutive years, using online surveys before and after its delivery and semi-structured interviews held after the course had finished. Analysis of pre-service teachers' responses revealed that their understanding of the differences between numeracy and mathematics improved after participating in the course, as did their confidence in incorporating numeracy into their teaching. It may be that the introduction of numeracy standards for graduate teachers increased the motivation of pre-service teachers and teacher educators alike to engage with numeracy across the curriculum as part of their respective professional obligations.

5 A Research Study: Teacher Preparation for Mathematical Literacy across the Curriculum in Ireland

The contribution of this chapter to research and development on mathematical literacy in initial teacher education complements previous studies, such as the examples presented above, each of which described and evaluated a single course or course component offered by one university in the Australian context. The aim of the research presented in the following sections was to investigate the preparation of secondary pre-service teachers in three Irish universities for their role in promoting mathematical literacy across the school curriculum and their ability to engage in mathematical literacy practices. For the purposes of this chapter, the analysis looks at pre-service teachers' understanding of numeracy, how they would embed numeracy practices in their subject specialism, and their opportunities to learn about numeracy in their ITE program.

5.1 *Research Design and Methods*

The study took a mixed methods approach, collecting quantitative and qualitative data from 204 pre-service teachers over the course of the 2018–2019 academic year. The participants were undertaking a Professional Master of Education (PME) degree, which is a two-year, postgraduate program for students who already have an undergraduate degree in a subject area designed to qualify secondary school teachers. Undergraduate teacher education programs are also available in Ireland, but the decision was made to sample PME programs because this is the most common pathway into secondary school teaching, and it offers the widest range of teaching subject specialisms. A secondary PME program was offered in all seven universities that existed in Ireland at the time of the study. All pre-service teachers who were enrolled in secondary PME degrees in three of these universities – designated University A, B, and C – were invited to participate in the study. The three universities were selected on the basis of their proximity to the location of the researcher (first author), thus facilitating travel to each site for data collection. University A and University B introduced numeracy as part of other courses within their initial teacher education programs, while University C had a semester-long course dedicated to literacy and numeracy for teaching.

The 204 pre-service teachers who agreed to participate represented 62% of the overall PME cohort in the three universities. Nineteen of these pre-service teachers attended University A, 82 attended University B, and 103 attended University C. Just over 60% of the participants were under 25 years of age. Their subject specialisms, categorised into curriculum groups, are shown in Table 2.2.

Between late September and early October 2018, pre-service teachers in each university were invited to complete a paper-based questionnaire during, or immediately after, a general pedagogy lecture that was attended by all PME students in their respective cohorts. The questionnaire consisted of 41 closed and open items. Its first section collected demographic data, while the second section assessed pre-service teachers' interpretations of numeracy and perceptions of their preparedness to teach for numeracy. A third section presented seven numeracy tasks for the pre-service teachers to solve. This chapter draws on participant demographic data and analysis of responses to the following three open-ended items from the second questionnaire section:

- What is your understanding of the term numeracy?
- Explain with the aid of a specific example how you would incorporate numeracy teaching and learning into a lesson.
- Do you think you have studied enough mathematics to be a competent teacher of numeracy? Explain your answer.

TABLE 2.2 Subject specialisms of pre-service teachers

Group	Subjects	Total (%)
STEM	Science (9), Technology (1), Mathematics (25)	35 (17%)
Social sciences	History (10), Geography (22), Business Studies (17)	49 (24%)
Arts and practical subjects	Music (12), Art (10), Physical Education (2)	24 (12%)
Languages	English (56), Languages other than English (39)	95 (47%)

Note: One participant left this question blank.

Questionnaire respondents were asked to provide their contact details if they were willing to participate in follow-up interviews, and twelve pre-service teachers did so. Five were from University A, two were from University B, and five were from University C. They were preparing to specialise in a wide range of subjects, including mathematics, music, English, languages, geography, and business studies. Two rounds of focus group interviews were planned in the first half of 2019, by which time all pre-service teachers had been introduced to some aspects of numeracy in their initial teacher education programs. The first focus group, in early January 2019, was to explore participants' understanding of numeracy and the second, in early May the same year, aimed to investigate their experiences in the initial teacher education program and their preparedness to teach for numeracy within their subject specialism. However, due to scheduling difficulties in Universities A and B, it was not possible to conduct the second round of focus group interviews and so individual telephone interviews were conducted instead with these participants. Interviews were audio-recorded and transcribed in full.

Thematic analysis (Braun & Clarke, 2006) was used to generate initial codes and then identify overarching themes from the open-ended questionnaire responses and interview transcripts, in order to answer the main research question that guided the study: How well-prepared are pre-service secondary teachers to teach for numeracy in subjects across the school curriculum in Ireland? Explanation of the thematic analysis process is presented in Table 2.3.

In this chapter, interest centres on pre-service teacher perceptions of their preparedness in terms of their understanding of numeracy, their descriptions of how they would incorporate numeracy into a lesson in their subject specialism, and the opportunities they experienced within their initial teacher education programs to learn about teaching strategies for numeracy across the curriculum.

5.2 *Pre-service Teachers' Understanding of the Meaning of Numeracy*

Thematic analysis of responses to the first questionnaire item yielded three distinct themes for describing the meaning of numeracy. A little more than half the pre-service teachers (113, 55.4%) described numeracy as a form of *mathematical knowledge*, for example, by referring to "numbers", "the ability to count", "maths skills", or "using figures, sequences or graphs". Just under one-third (64, 31.4%) interpreted numeracy as *using mathematical knowledge in different contexts*, mentioning that numeracy was "the study of maths in real life", or "using numbers in other subjects". A smaller proportion of respondents (20, 9.8%) explained numeracy in terms that referred to *thinking processes*, such as "decision making", "problem solving", and "logical". Seven participants left this question blank.

TABLE 2.3 Phases of thematic analysis

Phase	Explanation
Familiarising oneself with the data	All responses were read and notes were made of particular words/themes that emerged from the data.
Generating initial codes	Initial codes for the themes that emerged from the data were generated. Initial codes such as "Numbers" "Counting" "real life application of mathematics" "Problem solving" were generated from the initial analysis.
Searching for themes	The initial themes that were generated also corresponded with some elements of the numeracy model presented in Figure 2.1.
Reviewing themes	The sample of answers from the completed questionnaires were discussed by the researchers to ensure that the themes were reflecting the data appropriately.
Defining and naming themes	Each of the themes was re-named and coded. Some themes were amalgamated to reflect broader themes that emerged from the data.
Producing the report	Sample answers from the focus groups and questionnaires were selected that highlighted the theme that emerged from the data.

Further insights into pre-service teachers' understanding of numeracy were gained in the later focus group and individual interviews conducted with twelve volunteer participants from the three universities. By this time, all had been exposed to some aspects of numeracy for teaching, but only those from University C had experienced a formal course that addressed this topic. Pre-service teachers from University A claimed to understand what numeracy meant in their teaching specialisms; however, their understanding was limited to seeing numeracy as dates or numbers. For example, one female participant who specialised in teaching Irish language referred to:

> dates, years, and how to count … there's a variety of ways of counting things in Irish. (A3; F; Languages)

Those from University B said they did not understand what numeracy meant – a situation some also said they had noticed with practising teachers:

> Well, I don't really know if the truth be told, I'm trying to understand it myself. We had one lecture on it and it was literacy and numeracy about this time last year. But, other than that, it was glossed over. I really don't think there's a lot of understanding in numeracy in established teachers. (B1; M; Mathematics)

Participants from University C indicated that their understanding of numeracy had changed as a result of completing the semester-long course in literacy and numeracy for teaching. One male pre-service teacher, specialising in Business Studies and Geography, explained that before completing the course:

> It was just figures, the numeracy section in my [lesson] plans always had the date, just so that the students would know what date it is. (C4; M; Business Studies & Geography)

He went on to explain that the literacy and numeracy course:

> opened up my eyes to what numeracy actually is and the importance of context in numeracy as well. Let's say the percentages on loans or whatever, but then using that in context in the real world or real life settings. (C4; M; Business Studies & Geography)

5.3 *Pre-service Teachers' Perceptions of Incorporating Numeracy into a Lesson*

Responses to this questionnaire item described lesson ideas that were consistent with the pre-service teachers' interpretation of numeracy as involving either mathematical knowledge such as dates, time, percentages, units of measure; use of mathematical knowledge in real-world contexts such as finance; or thinking processes related to problem solving. A second layer of analysis classified the responses in terms of the level of curricular awareness displayed by each respondent; in other words, how well they could recognise the inherent numeracy demands of the subjects they were preparing to teach. The most common responses (82 pre-service teachers, or 40.2%) were classified as demonstrating *no curricular awareness* and included superficial answers such as "write the date on the board" or "listing the three tasks that students must undertake". The remaining responses were nearly evenly divided between those displaying *some curricular awareness* (55 responses, or 27.0%) and *relevant curricular awareness* (59 responses, or 28.9%). The former category included examples such as using statistics, reading and interpreting graphs, and calculating percentages, but without specific explanation of how these activities would help students learn the target subject. Examples of relevant curriculum awareness, which linked numeracy with subject-specific content and goals, included using ratio and proportion when introducing perspective in art, interpreting patterns in musical notation and scales, and investigating time-distance relationships in movement activities in physical education.

Interview responses largely confirmed these low levels of curricular awareness of the inherent numeracy demands within secondary school subjects, with pre-service teachers from Universities A and B usually mentioning trivial uses of numbers and dates in lessons. For example, a female pre-service language teacher from University B said:

> There were a few dates and I said, "How long?" and "What's the difference between the dates?" And I will consider it to be numeracy. (B2; F; Languages)

Those from University C seemed to have somewhat more curricular awareness of numeracy demands in the subjects they were preparing to teach; for example, a male pre-service teacher specialising in Geography and English explained:

> Overall, in terms of Geography, I do use a lot of graphs. Or even just speech writing in English, I do think using the statistics and discussing them is great. (C3; M; Geography & English)

5.4 *Pre-service Teachers' Opportunities to Learn about Numeracy for Teaching*

Two data sources provide insights into these future teachers' opportunity to learn about numeracy for teaching across the curriculum. First, one questionnaire item asked them to indicate whether they thought they had studied enough mathematics to be a competent teacher of numeracy. We acknowledge that mathematical knowledge is but one element of numeracy and other qualities and experiences are needed for teachers to become confident in recognising and responding to the numeracy demands of the subjects in which they specialise (Goos et al., 2014). However, previous research has found that pre-service teachers may be anxious about their personal mathematical knowledge and that this can hinder their engagement with courses designed to develop numeracy knowledge for teaching (Groves, 2001). There were 203 questionnaire participants who answered this question. Just under two-thirds (124, 61.0%) reported that they had studied enough mathematics to enable them to embed numeracy practices into their teaching; 46 (22.7%) indicated they had not done so; and 33 (16.3%) were unsure.

A statistical test was carried out to check for an association between pre-service teachers' subject specialism (using the groupings shown in Table 2.2) and their perception of mathematical preparedness to teach for numeracy. A chi-square test would usually be used for this purpose; but the assumptions underlying this test were violated because 31.3% of cells had expected counts of less than 5. A Fisher's exact text was conducted instead and revealed a statistically significant association (Fisher's exact test statistic = 30.956, $p < 0.0001$). A cross tabulation was then carried out to identify the nature and direction of differences between subject disciplines, as displayed in Table 2.4. Inspection of this table reveals that almost all pre-service teachers who were preparing to teach in the STEM disciplines (subjects such as Science, Technology, Engineering or Mathematics) (33, 94.3%) perceived that they had adequate mathematical knowledge for numeracy teaching across the curriculum – much more so than those in other subject specialisms, where only around half of respondents judged their mathematical knowledge to be adequate. There was also a high level of uncertainty amongst future teachers in the arts and practical subjects, almost all of whom were preparing to teach music or art, with 37.5% indicating they were "unsure" whether they had studied enough mathematics to be a competent teacher of numeracy. These findings suggest that initial teacher education programs should focus not only on curricular and pedagogical aspects of numeracy for teaching, but also boosting pre-service teachers' confidence in their own ability to promote numeracy across the curriculum.

TABLE 2.4 Cross tabulation of pre-service teachers' subject specialism against perception of adequacy of personal mathematical knowledge

	Subject specialism group			
Adequacy of personal mathematical knowledge	**STEM**	**Social sciences**	**Arts and practical subjects**	**Languages**
Yes	33 (94.3%)	26 (53.1%)	11 (45.8%)	54 (56.8%)
No	1 (2.9%)	13 (26.5%)	4 (16.7%)	27 (28.4%)
Unsure	1 (2.9%)	9 (18.4%)	9 (37.5%)	14 (14.7%)
Total	35 (100.1%)	49 (98.0%)	24 (100.0%)	95 (99.9%)

Pre-service teachers can also experience opportunity to learn about numeracy for teaching by studying this topic in their initial teacher education programs. In the Teacher Education and Development Study in Mathematics, Tatto et al. (2012) designed a questionnaire that asked future teachers who were about to graduate whether they had experienced opportunity to learn in several broad areas hypothesised to influence knowledge for teaching mathematics: tertiary-level mathematics, school-level mathematics, mathematics education pedagogy, general pedagogy, teaching diverse students, learning through school-based experiences, and the coherence of their teacher education program. In the present study, opportunity to learn about the meaning of numeracy and to develop subject-specific numeracy pedagogy was investigated by comparing the courses that pre-service teachers were required to study and their perceptions of the adequacy of these courses. As this issue was explored only in focus group and individual interviews involving twelve participants, any findings are illustrative rather than conclusive. Table 2.5 presents a description of what is offered in each of the University's ITE programs in relation to numeracy learning opportunities.

Pre-service teachers from University A and University B referred to their experiences in their initial teacher education programmes as "tick the box" exercises and "slightly confusing", as they felt they did not know enough about

TABLE 2.5 Information about courses offered by universities which incorporate numeracy

University	Name of course	Description of course
University A	Becoming a teacher: Identity and agency	Helps student teachers to identify and critically analyse influences which shape the individual in becoming a teacher and to focus on the key concepts of communication and reflection as significant processes in professional identity formation.
University B	Professional studies	Examines the role of assessment, curriculum, literacy and numeracy and ICT in education.
University C	Literacy and numeracy development in the post-primary classroom	Examines and reflects on the role, function, development and pedagogy of the core curriculum skills of literacy and numeracy in the post-primary classroom.

numeracy. For example, a female pre-service teacher at University A who was specialising in music and Irish language teaching described her experience in her subject specialism pedagogy lectures in terms that convey a superficial interpretation of numeracy:

> Yeah, we just had two hours. Now, in fairness, our Irish lecturer did say, she was just teaching us how to tick a box like, which is fair, we're here to get marks at the end of the day. But she did say, these are examples of ways you can include numeracy in your lesson planning: the date, the birthdays in your classes, what age are the students. I was using a date every single day, it didn't matter that they were giving me the exact same month, at least they were giving me the date. (A4; F; Music & Irish)

On the other hand, a male pre-service mathematics teacher from University B described receiving little or no guidance at all as to how to incorporate numeracy into his lessons, even though he was aware that he should be doing this:

> It's mentioned and stuff, like even in the lesson plans that they give us, there is a section for literacy and numeracy but like, we've never, never been really told what we should put in there. (B1; M; Mathematics)

In contrast, pre-service teachers from University C said they found the literacy and numeracy course they undertook to be very helpful. Nevertheless, they identified a missing connection as numeracy was never addressed in their subject specialism pedagogy courses where they were learning how to plan their lessons. For example, one interviewee explained:

> They could even mention it in the pedagogy classes. I suppose they don't mention it when we are in English pedagogy or learning to teach geography, you know. They don't mention numeracy at all, it's more literacy really. And they could definitely mention it in those lectures. (C5; M; English & Geography)

While this is only a very small interview sample, it is possible to hypothesise that a dedicated course on numeracy across the curriculum, even if taught in conjunction with literacy across the curriculum, is more effective for raising pre-service teachers' awareness of the numeracy demands of the subjects they teach than attempts to address numeracy within general pedagogy lectures or subject specialism pedagogy courses. This is an issue that needs to be investigated in further research.

6 Implications for Initial Teacher Education

It is perhaps a more challenging task to operationalise mathematical literacy in ways that make sense to teachers and that can be used by them for curriculum planning and task design across all subjects in primary and secondary schools. These curriculum mandates are sometimes accompanied by policies affecting the design and accreditation of initial teacher education programs, requiring graduating teachers to understand the meaning of mathematical literacy and to be able to implement teaching strategies that promote their students' mathematical literacy capabilities. However, there has been little research to date on how initial teacher education providers are addressing this requirement.

This chapter describes some examples of numeracy courses that have been designed, implemented, and evaluated in Australian teacher education programs as the background to a research study conducted in Ireland. The study, carried out in three Irish universities, investigated pre-service teachers' perceptions of their preparedness to develop their students' numeracy within their secondary school subject specialisms. Although most pre-service teachers felt that they had enough mathematical knowledge to become a competent teacher of numeracy – especially those specialising in the STEM subjects – they demonstrated largely superficial understanding of the meaning of numeracy in real-world and curriculum contexts and described mostly trivial examples of how they would incorporate numeracy into a lesson. Thus, there seems to be a lack of alignment between initial teacher education program requirements regarding numeracy and the pre-service teachers' lived experiences in their respective universities' programs.

Nevertheless, there were noticeable differences between the perspectives of pre-service teachers in Universities A, B, and C, which suggests that the concept of *opportunity to learn* might be a useful way of exploring these differences. Following Tatto et al. (2012) in the TEDS-M study, we define opportunity to learn in teacher education programs in terms of exposure to specific topics, that is, instructional content and time. Only in University C was there a dedicated course on literacy and numeracy in the school curriculum; in Universities A and B, numeracy was meant to be addressed as a topic within other courses focusing on pedagogy. Clearly, the greater allocation of content and time for numeracy in University C would have provided greater opportunity to learn than in Universities A and B.

Typically, there is not a lot of space in initial teacher education programs to satisfy all externally imposed accreditation requirements for program content, duration, and design (see AITSL, 2011; The Teaching Council, 2020). This makes it important to carefully consider alternative approaches to addressing numeracy as a required area of study. Four such approaches emerged from the literature and empirical evidence in this chapter:

1. A dedicated, semester-long course on numeracy across the curriculum;
2. A semester-long course that combines literacy and numeracy across the curriculum;
3. Some lectures on numeracy across the curriculum as a topic within a general pedagogy course;
4. Incidental attention to numeracy within subject-specific pedagogy (or "methods") courses.

Arguably, none of these options, on its own, will be sufficient, and an ideal approach might combine options 1 or 2 with option 4 so that pre-service teachers experience a thorough, research-informed grounding in numeracy alongside opportunities to apply their understanding of numeracy to develop resources and lesson plans in their subject specialisms.

Yet this recommendation raises another important question: Who should be responsible for providing opportunities to learn about numeracy across the curriculum? If numeracy is every teacher's responsibility, can the same kind of responsibility be placed on teacher educators? The numeracy activities and courses designed by Watson and Moritz (2002), Groves (2001), and Forgasz and Hall (2019) and described earlier in this chapter were led by mathematics teacher educators, but mostly in co-operation or consultation with colleagues in other curriculum areas. As Groves admitted:

> the mathematics education staff were acutely aware of their own ignorance of details of the curriculum in other subject areas, the numeracy demands and opportunities inherent in these areas, and, most of all, what actually happens in the classrooms and the teaching strategies which can be used to discern and respond to individual students' numeracy learning needs within these areas. (p. 52)

This observation suggests that a collaborative model of course design would be beneficial, led by mathematics education staff with expertise in numeracy across the curriculum but also involving teacher educators in other areas of the school curriculum. Such an approach would not only strengthen the content of a numeracy across the curriculum course, but also provide non-mathematics teacher educators with valuable insights into the inherent numeracy demands of their own subject specialisms that could then be shared with their teacher education students. In this way, teacher educators may experience opportunities to learn about numeracy across the curriculum that are just as worthwhile as the opportunities they offer to their students.

Acknowledgments

This chapter builds on the first author's doctoral thesis, supervised by Niamh O'Meara, Merrilyn Goos, and Paul Conway. Their contributions to the ideas discussed in this chapter are gratefully acknowledged.

References

Alberta Education. (2019). *What is numeracy?* https://education.alberta.ca/literacy-and-numeracy/numeracy/everyone/what-is-numeracy/

Australian Curriculum, Assessment and Reporting Authority (ACARA). (n.d.). *General capabilities*. https://www.australiancurriculum.edu.au/f-10-curriculum/general-capabilities/

Australian Institute for Teaching and School Leadership (AITSL). (2011). *Australian professional standards for teachers*. https://www.aitsl.edu.au/teach/standards

Beaton, A. E., Mullis, V. S, Martin, M. O., Gonzalez, E. J., Kelly, D. L., & Smith, T. A. (1996). *Mathematics achievement in the middle school years: IEA's third international mathematics and science study*. TIMSS International Study Centre. https://eric.ed.gov/?id=ED406419

Bolstad, O. H. (2020). Secondary teachers' operationalisation of mathematical literacy. *European Journal of Science and Mathematics Education, 8*(3), 115–135. https://doi.org/10.30935/scimath/9551

Braun, V., & Clarke, V. (2006). Using thematic analysis in psychology. *Qualitative Research in Psychology, 3*(2), 77–101. https://doi.org/10.1191/1478088706qp063oa

Connolly, C., Carr, E., & Knox, S. (2021). Diving deep into numeracy, cross-curricular professional development. *International Journal of Mathematical Education in Science and Technology, 54*(6), 1034–1053. https://doi.org/10.1080/0020739X.2021.1986160

Council of Australian Governments (COAG). (2008). *National numeracy review report*. Commonwealth of Australia. http://hdl.voced.edu.au/10707/143183

Department of Education and Skills (DES). (2011). *Literacy and numeracy learning for life: The national strategy to improve literacy and numeracy among children and young people 2011–2020*. https://www.education.ie/en/Publications/Policy-Reports/lit_num_strategy_full.pdf

Forgasz, H., & Hall, J. (2019). Learning about numeracy: The impact of a compulsory unit on pre-service teachers' understandings and beliefs. *Australian Journal of Teacher Education, 44*(2). http://dx.doi.org/10.14221/ajte.2018v44n2.2

Geiger, V., Goos, M., & Forgasz, H. (2015). A rich interpretation of numeracy for the 21st century: A survey of the state of the field. *ZDM Mathematics Education, 47*, 531–548. https://doi.org/10.1007/s11858-015-0708-1

Goos, M. (2018). The mathematics learning area: Conforming, reforming, or transforming? In A. Reid & D. Price (Eds.), *The Australian curriculum: Promises, problems, and possibilities* (pp. 55–64). Australian Curriculum Studies Association.

Goos, M. (2020). Mathematics crossing borders: A comparative analysis of models for integrating mathematics with other disciplines in pre-service teacher education. In N. Radakovic & L. Jao (Eds.), *Borders in mathematics pre-service teacher education* (pp. 91–115). Springer. https://doi.org/10.1007/978-3-030-44292-7_5

Goos, M. (2021). Numeracy across the curriculum as a stimulus for interdisciplinary inquiry in elementary school mathematics. In J. Novotna & H. Moraova (Eds.), *Broadening experiences in elementary school mathematics*. Proceedings of the 16th International Symposium on Elementary Mathematics Teaching (pp. 21–32). Charles University. https://www.semt.cz/

Goos, M., Dole, S., & Geiger, V. (2012). Auditing the numeracy demands of the Australian Curriculum. In J. Dindyal, L. Chen, & S. F. Ng (Eds.), *Mathematics education: Expanding horizons*. Proceedings of the 35th annual conference of the Mathematics Education Research Group of Australasia (pp. 314–321). MERGA. https://files.eric.ed.gov/fulltext/ED573240.pdf

Goos, M., Geiger, V., & Dole, S. (2014). Transforming professional practice in numeracy teaching. In Y. Li, E. Silver, & S. Li (Eds.), *Transforming mathematics instruction: Multiple approaches and practices* (pp. 81–102). Springer. https://doi.org/10.1007/978-3-319-04993-9_6

Goos, M., Geiger, V., Dole, S., Forgasz, H., & Bennison, A. (2019). *Numeracy across the curriculum: Research-based strategies for enhancing teaching and learning*. Routledge.

Goos, M., & O'Sullivan, K. (2023). The evolution and uptake of numeracy and mathematical literacy as drivers for curriculum reform. In Y. Shimizu, R. Vithal (Eds.), *Mathematics curriculum reforms around the world. New ICMI study series*. Springer. https://doi.org/10.1007/978-3-031-13548-4_21

Groves, S. (2001). Numeracy across the curriculum: Recognising and responding to the demands and numeracy opportunities inherent in secondary teaching. *Mathematics Teacher Education and Development, 3*, 48–61. https://mted.merga.net.au/index.php/mted/article/view/122

Liljedahl, P. (2015). Numeracy task design: A case of changing mathematics teaching practice. *ZDM Mathematics Education, 47*, 625–637. https://doi.org/10.1007/s11858-015-0703-6

Ministry of Education. (1959). *15 to 18: A report of the Central Advisory Council for Education*. HMSO. http://www.educationengland.org.uk/documents/crowther/crowther1959-1.html

National Council for Curriculum and Assessment. (2021). *Framework for junior cycle*. https://ncca.ie/en/junior-cycle/framework-for-junior-cycle/

Niss, M. (2018). National and international curricular use of the competency-based Danish "KOM project". In Y. Shimizu & R. Vithal (Eds.), *ICMI study 24 Conference Proceedings. School mathematics curriculum reforms: Challenges, changes and opportunities* (pp. 69–76). University of Tsukuba & ICMI. http://www.human.tsukuba.ac.jp/~icmi24/

Norwegian Directorate for Education and Training. (2012). *Framework for basic skills.* https://www.udir.no/in-english/Framework-for-Basic-Skills/

Organisation for Economic Co-operation and Development (OECD). (2018). *PISA 2021 mathematics framework (draft).* https://www.oecd.org/pisa/publications/pisa-2021-assessment-and-analytical-framework.htm

Shiel, G., Kelleher, C., McKeown, C., & Denner, S. (2016). *Future ready? The performance of 15-year-olds in Ireland on science, reading literacy and mathematics in PISA 2015.* Educational Research Centre. https://www.erc.ie/wp-content/uploads/2016/12/PISA2015_FutureReady.pdf

Steen, L. A. (Ed.). (2001). *Mathematics and democracy: The case for quantitative literacy.* National Council on Education and the Disciplines. https://d3n8a8pro7vhmx.cloudfront.net/typp/pages/67/attachments/original/1445572030/8952_MathAndDemocracy(2).pdf?1445572030

Tatto, M. T., Schwille, J., Senk, S., Ingvarson, L., Rowley, G., Peck, R., Bankov, K., Rodriguez, M., & Reckase, M. (2012). *Policy, practice, and readiness to teach primary and secondary mathematics in 17 countries. Findings from the IEA Teacher Education and Development Study in Mathematics (TEDS-M).* International Association for the Evaluation of Educational Achievement. https://eric.ed.gov/?id=ED542380

Tasmanian Government. (2019). *Numeracy in the news.* http://www.tas-education.org/numeracy/default.html

The Teaching Council. (2020). Céim: Standards for initial teacher education. https://www.teachingcouncil.ie/en/news-events/latest-news/ceim-standards-for-initial-teacher-education.pdf

Thornton, S., & Hogan, J. (2004). Orientations to numeracy: Teachers' confidence and disposition to use mathematics across the curriculum. In M. J. Hoines & A.B. Fuglestad (Eds.), *Proceedings of the 28th conference of the international group for the psychology of mathematics education* (Vol. 4, pp. 313–320). PME. https://files.eric.ed.gov/fulltext/ED489538.pdf

Watson, J., & Moritz, J. (2002). Quantitative literacy for pre-service teachers via the internet. *Mathematics Teacher Education and Development, 4*, 42–55. https://mted.merga.net.au/index.php/mted/article/view/115

CHAPTER 3

The Teaching of Mathematical Literacy in Four South African Classrooms

Ellamma Pillay and Sarah Bansilal

Abstract

In South Africa, the subject Mathematical Literacy (ML) was introduced as a separate subject, different in scope and purpose from the core mathematics, and reserved for those learners who did not study core mathematics in their last three years of their schooling. The purpose of this chapter is to explore the ways in which four South African ML teachers respond to the demands of mediating the subject with their learners. Data for the study were generated by classroom observations and interviews with the teachers. The study found that the two specialist mathematics teachers tended to foreground the mathematics rather than using the context to make sense of the problems. This may be because of their own belief systems that prioritized mathematics. The two non-specialist teachers struggled with the mathematics and the contextual concepts on many levels, and showed an over-reliance on the textbook. The results suggest that both specialist and non-specialist mathematics teachers need professional development opportunities that would help them better understand the purpose of ML. Teachers also need guidance in issues such as the integration of the content and context in teaching ML.

Keywords

mathematical literacy – real life context – South Africa – contextual attributes

1 Introduction

There is much attention in research studies about the mathematical literacy skills and competences that children need to develop so that they may be able to participate in the 21st century. There is also a growing awareness of the importance of mathematical literacy for personal empowerment and becoming a critical citizen (Hoogland, 2021). The Organization for Economic

 | DOI:10.1163/9789004710962_003

Cooperation and Development (OECD, 2018, p. 6) has defined mathematical literacy as:

> the capacity of individuals to reason mathematically and solve problems in a variety of 21st century contexts, [and to] use mathematical concepts, procedures, facts and tools to describe, explain and predict phenomena.

Many countries, including South Africa have engaged in curriculum revisions to enable the preparation of learners to become critical thinkers who can contribute to solutions to complex problems. A particular concern of the South African education department was that many learners opted not to study any form of mathematics in the last three years of their schooling (Grades 10–12). Consequently, in South Africa, Mathematical Literacy (ML) was introduced as a separate subject, different in scope and purpose from the symbolic and formal school mathematics subject to "enable the learner to become a self-managing person, a contributing worker and a participating citizen" (DBE, 2011, p. 8). ML learners are encouraged to explore the meaning behind numerical and graphical information in the media and elsewhere and to extrapolate how these can affect their lives. The intention is that when learners encounter these contexts in their current or future lives, they will be better able to make informed decisions, that is, the subject takes on a life-preparation perspective (Venkat, 2010, p. 55). The curriculum documents emphasise an interplay between content, contexts, and problem-solving skills that are presented as basic skills and application topics (Brombacher, 2007). The Basic Skills topics comprise: Interpreting and communicating answers and calculations; Numbers and calculations with numbers; and, Patterns, relationships and representations. The Application Topics comprise: Finance; Measurement; Maps, plans and other representations of the physical world; Data handling; and Probability.

With these changes to the school curriculum in South Africa, over a decade later many teachers and researchers are still trying to understand some of the complexities of teaching and learning of ML. It is important to note that the subject is taught mainly by mathematics specialists and specialists in other subjects, and there has been limited attention to in-service training of current ML teachers (Bansilal et al., 2012; Webb et al., 2013). Furthermore, the subject ML is studied by 15–18-year-olds who do not study core mathematics and are typically those who were not so successful in mathematics up to grade 9. Hence ML in South Africa is situated within a context that is different from others where mathematical literacy is considered to be part of the mathematics classroom, or integrated across the curriculum. In this chapter, we look at

the practices of four teachers as they mediated the subject with their learners. The research questions that guide this chapter are:
- How do the teachers negotiate the demands of mediating the subject ML with their learners?
- What insights can be gained from their approaches to the teaching of ML?

It is hoped that this study can add more knowledge about the demands of teaching and learning mathematical literacy irrespective of whether developing mathematical literacy is taken as the responsibility of the mathematics class or integrated across the curriculum. By drawing attention to some of the potential problems that may be encountered by teachers, it illustrates that teaching ML is different from the teaching of mathematics.

2 Literature

In recent years, there has been much interest in how we could improve school learners' competencies so that they are better prepared for life in the 21st century. The terms numeracy, statistical literacy, quantitative literacy as well as mathematical literacy are variously used in different countries to describe the desired numerical/mathematical knowledge and skills expected from students (Gal et al., 2020). Although there is not yet an agreed definition of the notion of mathematical literacy, there are some agreed components of mathematical literacy (Turner, 2021). These are practical aspects such as number sense, arithmetic, and spatial skills amongst others; procedural knowledge which refers to knowledge, procedures, theorems, definitions, etc. and a more pervasive way of thinking about mathematics such as thinking mathematically, reasoning, communication, etc. (Turner, 2021). The Common European Numeracy Framework (CENF) emphasizes dispositions, content knowledge and skills, context as well as higher-order skills in improving numerate behaviour (Hoogland, 2021). Savard and Cavalcante (2021) highlight the role of financial numeracy in developing citizenship competencies and call for greater prominence in the curriculum for financial numeracy. The authors advise that preservice teachers also need a course on financial numeracy which could focus on developing their knowledge and thinking about financial mathematics concepts as well as their skills in designing and adapting tasks for financial numeracy (Savard & Cavalcante, 2021; Sawatzki et al., 2024).

When ML was introduced in South Africa, there were no pre-service programmes to train ML teachers, and teachers who taught other subjects were encouraged to teach ML (Webb et al., 2013). Researchers wondered about whether teachers teaching ML would focus more on the mathematics appearing in the context-specific problem, or whether they would foreground the process of solving and interpreting the context-specific problem (Graven & Venkat, 2007). The inclusion of real-life contexts in the ML classroom requires mathematical reasoning together with other contextual considerations implying that "boundary crossing" between mathematical and contextual tools should be a critical feature of ML teaching (Venkat & Winter, 2015). The boundary is marked "by mathematical activity and conventions on one side, and contextual activity and conventions on the other side) (Venkat & Winter, 2015, p. 576).

In recognising that tasks themselves differ in the degree to which the context is emphasized, Almuna (2016) summarized three levels of context use within tasks. A zero-order use of context is when the task just requires mathematics terms, symbols, and translation of textual mathematical information (such as calculating the cost of seven pencils if the price of one is given). A first-order use of context is when the context is needed to judge the suitability of the answer for the problem (for example, when one is asked to work out the transfer cost fee for a house, given the price and the transfer duty rule). A second-order context is when "students need to move backwards and forwards" between the problem and the context to set up the problem and to reflect on the suitability of the answer (Almuna, 2016, p. 112). With the second-order use of context, Almuna (2016) notes that information needs to be drawn from knowledge of the context itself not only from the problem description (e.g., calculating what one's current salary would have been if the annual increase had kept up with inflation in the last three years). The curriculum descriptions for ML (DBE, 2011) prioritise the use of second-order contexts in ML classrooms, by stipulating that mathematics content should be taught in an integrated manner and should "not be taught in the absence of context" (DBE, 2011, p. 8). However national assessments and observations of classroom practice show that this is not the case in practice (North & Christiansen, 2015; Pillay, 2018).

Some studies have focused on how learners and teachers deal with the demands of ML tasks (Bansilal, 2017; Bansilal & Debba, 2012; Bansilal et al., 2012). Findings show that contextual language is specific to each task with its own demands and can present barriers to learners in understanding the context as well as understanding the instructions, particularly to those with poor English language skills (Debba & Bansilal, 2012). Going deeper into various rules used in the context, it was found that the contextual rules varied in demand according to the complexity of the language used to explain the rule, the amount of information required to apply the rule, as well as the degree of familiarity with the rule (Bansilal & Debba, 2012). The complex language embedded in the transfer duty rule: "For a purchase price of R500,001 to R1,000,000, the transfer duty is 5% on the value above R500,000", led to misinterpretations of the rule (Bansilal et al., 2012). Many teachers did not consider the last part of the statement and worked out the transfer duty of a house costing R895,000 as 5% of the full amount, instead of 5% of R395,000 (value above R500,000). One study that used theorems-in-action to identify the different interpretations of the contextual rule for inflation by 406 teachers, uncovered more than 20 different ways in which the relationships between the quantities were constructed (Bansilal, 2017). These studies indicate that engagement with and understanding of concepts in various contexts can be quite complex, even for teachers, and further research is needed about specific problems set within real-life contexts so that teachers can develop improved pedagogical strategies to help learners develop their mathematical literacy skills.

Many studies concur that teachers need support in integrating numeracy into the curriculum or developing mathematical literacy skills (Bolstad, 2021; Geiger, 2021; Webb et al., 2013). Bolstad (2021) notes that curriculum documents do not provide sufficient direction on their own for teachers to operationalise the different demands and opportunities for teaching mathematical literacy. Teachers who teach in this area need support in improving their mathematics understanding; developing a clear understanding of what mathematical literacy is; working out how to integrate financial numeracy into the curriculum; learning to use authentic situations to develop tasks that serve as a basis for promoting students' numeracy; and, in adopting investigative pedagogies to fully realise the numeracy opportunities available in a range of contexts (Bolstad, 2021; Goos et al., 2014; Savard & Cavalcante, 2021; Turner, 2021; Webb et al., 2013). Goos et al. (2014) found that it was important to have a clear model of numeracy outlining essential elements to enable changes in teachers' numeracy practices. Without a clear understanding of how such tasks can be created or adapted, teachers find it difficult to develop suitable activities for their classrooms. In describing the approaches of four ML teachers in South Africa, it is hoped that these cases can illustrate some of the issues that emerge in classrooms where teachers try to develop mathematical literacy skills in general.

3 Framework

Our perspective is that ML is a subject that entails participation within mathematical conceptual and contextual domains, to solve problems. Greeno (1991) describes a conceptual domain in mathematics as an environment with tools and resources at various places in the domain. Knowing the domain means participating in the domain and being able to recognise, find and use those resources productively (Greeno, 1991). Tools and resources for the mathematics domain refer to concepts, procedures, symbols, and representations that are required for participation.

Noting that a context is the frame for the event or phenomenon being examined (Duranti & Goodwin, 1992), a contextual domain in ML can be seen as an environment with its own tools and resources. Drawing on previous work (Bansilal & Debba, 2012), we identify three contextual attributes that can provide the tools or resources for educational contexts:

Contextual concepts are objects used in the context to convey specific information and whose meaning is bound to the context itself. An example of contextual concepts is inflation which refers to an overall increase in the Consumer Price Index (CPI), which is a weighted average of prices for different goods. Another example is that of fuel consumption of a car.

Contextual language refers to the ways "in which talk itself invokes context and provides context for other talk" (Duranti & Goodwin, 1992, p. 7). In this study we use the phrase contextual language to refer to words or phrases which hold a particular meaning within the context. For example, "200 free kilometres per day" in the car hire scenarios refers to the situation where the contract allows one to drive up to 200km a day without incurring additional charges. A second example is that of a fee that is described as a certain amount per hour or 'part thereof' which implies that even if the person works for part of an hour, the charge will be for the full hour.

Contextual rules are bound to the context that need to be applied and interpreted by the learner. For example, the rule used to calculate the inflation rate is given by: inflation rate $= \frac{p_1 - p_0}{p_0} \times 100\%$, where P_1 and P_0 refer to the prices of goods in the current and the previous year respectively. A second example taken from this study is that of volume of water collected = surface area of roof in $m^2 \times$ rainfall in metres.

The elaboration of these contextual tools and resources provides a language for us to describe the teachers' approaches and interactions in their ML classrooms.

4 Methodology

This qualitative study formed part of the doctoral study by the first author with the purpose of exploring how teachers mediate ML concepts with their learners. In this chapter, we consider the data from four teachers Kelan, Danny,

Edith, and Grace. The teachers were selected using convenience sampling while making sure that there was a mix between those who had and did not have specialist mathematical training. Data were generated by three classroom observations and an interview with each teacher. The observations were conducted by the first author, who generated field notes during the observations. After an initial analysis of the observation, the interview was conducted using the video stimulated recall technique (Abdulhamid, 2022). The teachers were probed about their perspectives in general about teaching ML and more particularly, about particular aspects that emerged during the observation. The interviews were video-recorded and then transcribed. We also conducted a document analysis based on the teaching materials used by the teachers, before the interview.

The data were synthesised into four narratives, one for each teacher, using Polkinghorne's (1995) narrative analysis technique. In this type of analysis, the researcher's task is to configure the data elements comprising the observation notes, interview transcripts, teaching materials, and the field notes of the first author into a story. The intention of this process is to give meaning to the data as contributors to the goal of identifying issues related to the teaching of, and for mathematical literacy. The narratives were used to provide insight into how ML teachers negotiated the demands of developing mathematical literacy competencies with their learners. The resulting narratives were studied by both authors to ensure that they reflected the research findings accurately.

5 Results

The results are presented as narratives of each of the four teachers with subheadings that serve as descriptors of the teachers' approaches to the teaching of ML.

5.1 *Kelan*

Kelan is male and the head of the mathematics department at his school, he has taught mathematics for 25 years. Kelan often took the opportunity to make links between and emphasise underlying mathematics concepts.

5.1.1 Foregrounding Mathematics Concepts

When discussing the different representations of the function of the cost versus the number of sweets bought, he brought in the concept of dependent and independent variables. As he worked with each of the three representations (formula, table, graph) he used the opportunity to point out conventions about using dependent and independent variables in each.

He similarly foregrounded the concepts of direct and indirect proportion. When working with the graph of the number of balloons that were shared equally amongst children, he explained that it was an example of indirect proportion. While discussing a task using the context of buying sweets, he reminded them about direct and indirect proportion:

> If we had zero sweets, it's going to cost me zero cents. So, that means this line is going to start at zero, okay ... All the dots will lie on the straight line. Now that shows a direct proportion, ok. All of them are increasing.

Kelan hence used the opportunity to explain that these points lie on a straight line that went through the origin and therefore represented a direct proportion. He revisited the terms during the discussion of another task based on the cost of buying "popsicles" which are frozen juice sachets. Note that K stands for Kelan.

> K: One popsicle costs R2. Now, if 10 friends were there, ... how much money should you have ...
>
> L: R20.
>
> K: R20, so now this has got a direct relationship. ... Now what type of proportion is this? Direct or indirect? ... What is direct? When one increases the other also increases. What is indirect? When one increases the other decreases.

In this conversation Kelan merely recalled the meaning of direct versus indirect proportion with learners, without engaging them further, thus missing an opportunity to distinguish more clearly between the two types of proportion.

5.1.2 Providing a Concrete Illustration of Area Covered by a Tin of Paint

Kelan in discussing a function between the number of tins and the area covered in m^2, emphasised that it was the area that was being covered by a tin of paint. He pointed to a rectangular portion of the chalkboard that was painted to illustrate the meaning of area:

> I can look at the amount of paint I could use to cover that area, okay. Now, the amount of paint that's needed to cover the 12 m^2. 12 m^2 generally would be the size of a rectangle that's 3 m wide and 4 m long, ... or ... 6 m long and 2 m wide.

Hence Kelan easily provided a concrete illustration of "area covered in m^2". However, Kelan was not so successful in trying to identify a metaphor to access the meaning of slope in a distance versus time graph, as discussed next.

5.1.3 Trying to Provide Access to a Mathematics Concept Using a Misleading Metaphor

Kelan tried to focus on the steepness of the gradients of a distance-time graph made up of three parts as shown in the task in Figure 3.1.

Kelan tried to lead his learners to understand that the gradient is steeper when the speed is greater, that is, when a greater distance is covered in a shorter time:

> K: Now, when you look at these three different lines, right ... so, when you look at the run the first time, and then when you look at the cycle. Now, what is the difference in these lines ... (points to the graph he drew on the board) when compared to this horizontal line here? ...
>
> L_1: That's not straight.

When his learners did not respond the way he expected, Kelan tried to direct their attention to one line being steeper than another, by referring to two roads:

> Right, now when you coming up on Malaya Drive, ok, right, and then when you coming up Siaadra Road (two roads coming up to the school), now what's the difference between ... (them)

The learners responded with answers such as: "different heights", "they go at an angle" or "got speed bumps – slows speed down" and when one student said "One is steeper than the other", Kelan was relieved and responded:

An athlete trains by running, cycling and then running again. The graph shows the distance she trains against time taken.

a) How far does she travel altogether?
b) How far does she cycle?
c) Which is done faster, the first or the second run?
d) What is the average speed of the first run in kilometres per minute?
e) How long did she train for altogether? Give your answer in hours and minutes.

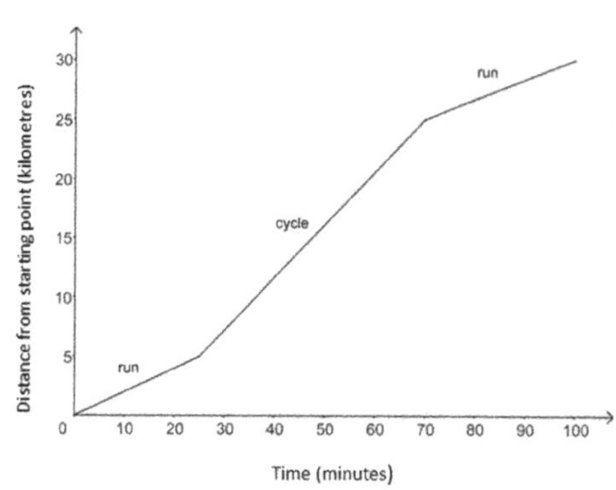

FIGURE 3.1 Task based on lines with different gradients (Source: Teacher's notes)

> Very good, excellent. One is steeper than the other ... Now, so what it means is this, that, when we run, we run slower than when we cycle. So therefore, you'll find that one line is steeper than the other. So, we talk about, what you call, gradient here.

Thus, Kelan finally drew attention to the notion of a steeper slope. However, in Kelan's real-life analogy of a steep road, the road with the steeper gradient meant that a person will travel at a slower speed, whereas the mathematical example illustrates the opposite- that in a graph of distance vs time, the line with a steeper slope illustrates a faster speed. This incident illustrates that finding suitable metaphors from a real-life context to illustrate mathematical concepts can sometimes be tricky.

5.2 *Danny*

Danny is a male teacher who has been teaching Mathematics for more than thirty years. Danny's teaching focus was on the examinations, choosing to first review the previous Grade 11 examination paper over two lessons, to help learners understand their mistakes and misconceptions, since they had performed poorly.

5.2.1 Moving Quickly to the Mathematics Calculations

Danny's approach was to use the context to introduce the problem and then ignore the context once the calculation steps were evident. The following extract from a task in Figure 3.2 shows this technique.

Danny drew a rectangle on the chalkboard to represent the surface area of 90 m^2 covered by the roof and said:

Question 4

4.5 If Danielle lives in KwaZulu-Natal, which receives about 845 mm of rain annually, and her house has a roof that covers $90m^2$, find the volume of water collected in litres.
(Volume of water collected = surface area of roof in m^2 x rainfall in metres.)
($1m^3$=1000 litres)

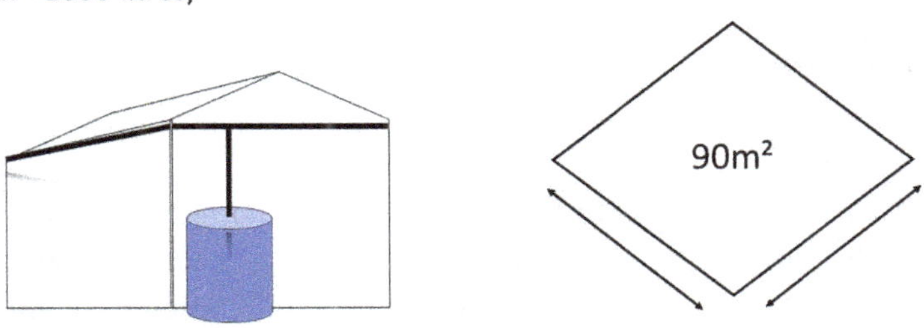

FIGURE 3.2 Question 4.5 focusing on the rainwater runoff from the roof (Reproduced from DBE, 2012, p. 6)

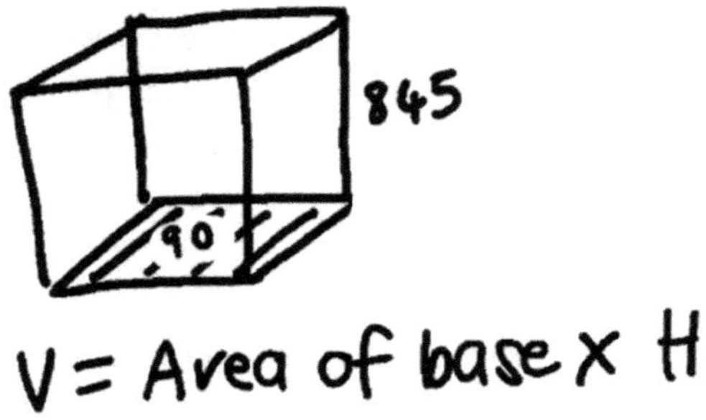

FIGURE 3.3
Volume of imaginary prism
(Source: Authors)

> Here's the volume, I raise it up here, your volume will be the space inside. Now the volume will be in this case, area of base times the height. ... Area of base times the height. Area is the bottom already given here times the height (Danny draws dots on the base of diagram and writes 90). Now it rained ... in Natal it rained 845 mm. Can you see? So that becomes the height. (writes 845 on the diagram)

While explaining, he drew in a height on the rectangle as shown in Figure 3.3, thereby converting it to a 3D rectangular prism. He marked the height as 845 mm (the annual rainfall), and wrote V= Area of base × H, thus guiding his learners on how they could utilise the contextual information in the given rule.

He did not spend time explaining why the contextual rule (area of roof in m^2 × rainfall in metres) could be represented as the volume of the imaginary prism, or how the annual rainfall could be used as the height. Danny thus quickly stripped off the context so that they could move to the calculation of the volume of a prism.

5.2.2 Emphasising Rules for Conversion of Units

After getting the volume in terms of m^3, Danny then focused on converting this to litres, pointing to the rule given in the instruction. Note D stands for Danny while L_1 stands for Learner 1 and so on.

> D: They tell you here, $1m^3$ is equivalent to 1000 and I now have, $76{,}05m^3$ is equal to? How do I find that?
>
> L_1: 76,05 divided by a 1000.
>
> D: You divide?
>
> L_2: 1000 ...
>
> D: You divide or you multiply it? One metre is that one. So I multiply. It's 76,05 × 1000. And what that will be? Moving decimal places from this, you move 3 [gesturing that the decimal comma was moved three places to the right], 76 050

Danny hence linked multiplication by a thousand, with moving the decimal comma to the right of the digits without trying to use the context to make sense of the conversion.

5.2.3 Unfolding Information in Steps to Reveal the Mathematical Structure

As seen in Figure 3.3, Danny drew the diagram of the rectangular prism, in stages to help learners see the link between the given formula for volume of a prism and the volume of the rain runoff from the roof. Danny explained his strategy:

> If you write something on the diagram, the child can also write it on his diagram. You make changes on the diagram, you break it up the child can do it there.

By drawing the diagram in stages, it allowed Danny to deal with the information a little at a time so that the learners were not bombarded with all the information at once.

5.3 *Edith*

Edith is a non-specialist teacher who specialised in language teaching and has been teaching ML for the past 7 years. In the three observation lessons, Edith worked with ratios and rates of different types, and drew her examples from a worksheet used for mathematics in an earlier grade.

5.3.1 Providing Rules to Make It Easier

Edith felt that it was important for ML learners to understand the basic mathematics content and she tried to show them rules to make it easier: "You have to know basic rules of maths, you have to teach them ground rules". However, Edith was not specific about which rule to use in which situation or why certain rules were used.

5.3.2 The "Cross-Multiplication" Rule

Edith highlighted the use of the cross-multiplication, when dealing with a missing value problem about the number of teachers required for 1178 learners, given the teacher-to-learner ratio of 1:38. Edith wrote the ratios teacher: learner and 1:38, one below the other. After some discussion, she asked "Where does x come? Do we know the number of teachers?" Edith then added the next line x:1178 as shown in Figure 3.4. She then explained "You cross multiply". Edith then drew in a cross and got to the next steps shown in Figure 3.4 to arrive at the answer of 31 teachers.

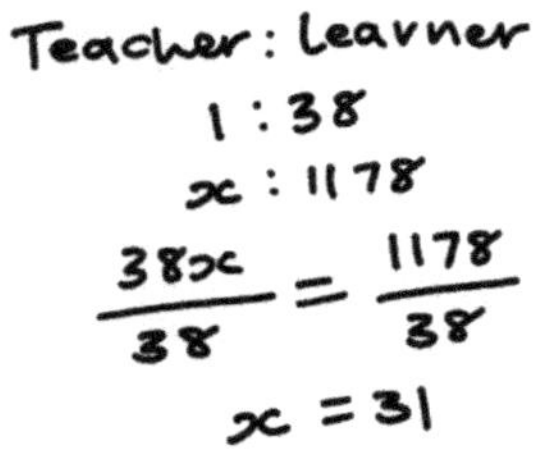

FIGURE 3.4
Ratio representation of the teacher-learner problem
(Source: Authors)

In her cross-multiplication approach, Edith went directly from writing the ratios to arriving at the equation $38x = 1178$. She did not explain the movement from the ratio form

$$1:38$$
$$x:1178$$

to the equivalent fraction form

$$\frac{1}{x} = \frac{38}{1178} \text{ or } \frac{1}{38} = \frac{x}{1178}$$

which provides the mathematical justification for the equation $38x = 1178$. Her approach implies that there is a special mathematical operation of cross multiplication that can be performed "across" the quantities making up the equivalent ratios, commonly seen in many classrooms (Bansilal & Ubah, 2020; Mahlaba & Bansilal, 2015). Edith also experienced difficulty in interpreting the contextual language embedded within a task as described next.

5.3.3 Difficulties in Interpreting Contextual Language

In a task focusing on fuel consumption (amount of fuel used to cover a certain distance), Edith avoided its meaning:

> Different cars have different levels of fuel consumption. That means different cars will have different petrol tanks, isn't that so? You gonna put fuel into the tanks.

Edith explained different fuel consumptions as something that seems to be associated with the difference in the size of fuel tanks, and in so doing, she side-stepped the actual meaning of the contextual term 'fuel consumption'.

This technique of sidestepping terms from the context was also evident in Edith's explanations about the phrase average rate. She started by announcing the term average rate, implying that she was going to discuss the meaning of average rate but then meandered through various issues: "Average rates. Now consider when we say that a car travelled at a speed of 120km per hour during a journey, clearly the car did not travel at precisely 120km".

She then digressed and commented that the speed of other cars affects one's own speed: "No, their travelling might be faster or might be slower. So you're decreasing your speed as you're moving on".

Edith then brought her own example of driving to school and why she could not drive at 120 km/h "like this morning, in my case, I came across a traffic jam. I can't travel at a 120km per hour to get to you people, because I had traffic".

Hence, she focused on why one cannot travel at a certain speed, and not an explanation of what constitutes average speed. Her discussion covered many issues except for the meaning of average rate.

5.4 *Grace*

Grace is a non-specialist teacher whose specialisation was in commercial subjects. She has been teaching ML for 6 years. In the first two lessons, Grace discussed various tasks based on linear relationships that were or were not directly proportional. In the third lesson, she discussed inverse proportional relationships.

5.4.1 Reliance on the Textbook

Grace worked very closely with the textbook and the teachers' guide which had solutions for the various tasks, but faltered when she tried to explain something that did not appear in the guide as in the case of the mathematics–specific phrases of direct and indirect proportion as well as continuous and discrete variable. In the interview she explained that the book was her guide about what she needed to explain the concepts: "Myself, I am being helped by the book. The book also gives you some clear lines what, what to do".

5.4.2 Difficulties in Explaining Differences between Two Different Mathematics Relationship

Grace tried to explain direct and indirect proportions to her class using her textbook as shown in the extract of Figure 3.5. Grace read:

> Okay, linear relationships that are not directly proportion. Because this one (previous example) is directly proportional, there is an increase, which with the cost per minute according to how many minutes you spend, so it is directly proportional. Now, if they are saying not directly proportional, you can say there is an inverse and opposite

Linear relationships that are not directly proportional
In some linear relationships the dependent variable has a fixed value at the start. Example: Where a cell phone contract has a basic fee of R95 plus a cost per minute charge, the dependent variable starts off at R95 (basic fee) and the independent variable is the amount of minutes used.

FIGURE 3.5 Extract from the textbook read by Grace (Reproduced from DBE, 2015)

Grace's explanation above revealed an error in her reasoning. Her assumption was that if a linear relationship was not a directly proportional one, then it would be an inversely proportional relationship. After presenting this explanation, Grace went back to read directly from the textbook about the topic "Linear relationships that are not directly proportional". However, the textbook did not provide any clear description of these kinds of linear relationships, and Grace went on with her reading:

> (Reading from textbook – announcing the topic) Linear relation that are not directly proportional … (continuing with the reading) In some linear relationship there is a dependent variables you know that here you are having a dependent and independent minutes and the cost which is the dependent and independent.

Grace then moved on to a discussion on another topic without addressing how directly proportional relationships fit into the set of linear relationships.

Grace similarly faltered when she tried to explain the terms discrete and continuous when discussing two tasks, one based on the calculation of an electrician's fee and one on the temperature of a freezer which has been switched on. The first task was based on an electrician's fee, where he charged "a call-out fee and then R150/hour or part thereof" (DBE, 2015, p. 124). Learners were asked to calculate the cost for various hours, by completing a table of the form given in Table 3.1.

They were also asked whether the relationship was increasing or decreasing and if the values were discrete or continuous. Grace categorised the values given for the electrician fee charge as discrete values: "They (the values) are

TABLE 3.1 Table of values that learners were asked to complete

Number of hours worked	1	2	3	4	5	6	7
Fee charged (R)	350	500					

discrete because they are having a difference of 150". Grace's reason was not complete. Normally a function such as $y = at + c$, where t is the time, would be continuous, however in this setting the context–specific phrase "part thereof" makes the function stepwise, so the dependent variable takes on a set of discrete values. This phrase means that if the electrician works for a part of an hour, he charges for a full hour, so the charges could be R350, R500, R650, etc. However, Grace did not engage with the phrase, and it is of interest to note that for this particular question, the teacher's guide had the answer (discrete) without giving a reason.

There was another time where she faltered when the textbook did not explain the term continuous, and Grace passed it off by saying: "It's continuous because it is decreasing".

5.4.3 Difficulties in Understanding a Context

The context for Task 2 is that of a freezer whose temperature decreases when the power is switched on. However, Grace misinterpreted this to mean that the temperature decreased because the freezer was switched off: "The temperature is decreasing after every hour. After every hour switch off the freezer, the freezer is off; the temperature is decreasing".

5.4.4 Difficulties in Understanding a Contextual Rule

The task on the electrician's fee specified a call-out fee and thereafter, R150 per hour or part thereof of the time he spent. When the time taken was more than 8 hours, there was a "20% discount on the hourly fee for the hours in excess of 8 hours" (DBE, 2015, p. 124).

Grace engaged with her learners and first worked out the cost for the first 8 hours, using R200 + R150 × (no of hours) and wrote these on the board – see Figure 3.6.

In working out the cost for 9 hours, she followed the same steps and got the answer of R1550. She then subtracted 20% of R1550 from the total to get R1240, as seen in Figure 3.6. Her work showed a misunderstanding of the phrase "hourly fee for hours in excess of eight hours". Instead of calculating the discount on the

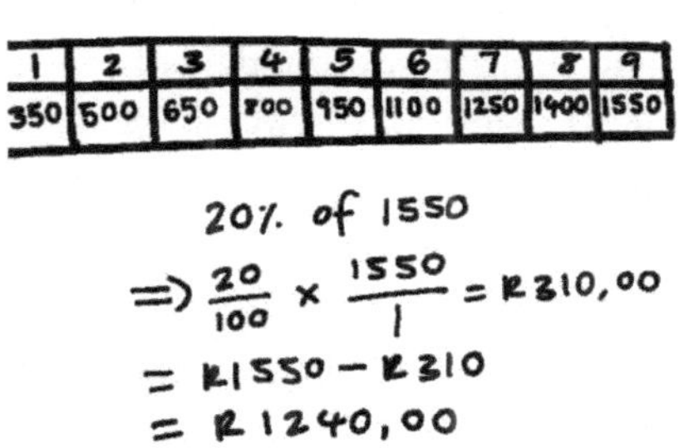

FIGURE 3.6
Calculations for Question 2.1 (Source: Authors)

hourly fee in excess of eight hours, Grace calculated the discount on the total amount charged for nine hours. It is interesting to note that neither Grace nor the learners queried that for seven hours, the total fee charge (R1250) was more than that for nine hours (R1240).

6 Discussion

The four cases presented in the study illustrate some of the issues related to the teaching of ML which are explored in further detail below.

The vignettes of the two specialist mathematics teachers showed that they often tried to foreground the mathematics in their lessons, even when discussing problems set within contexts. Graven and Venkat (2007) argued that ML teachers have different agendas that direct their approaches when working with tasks in the ML classroom. In this setting, it was evident that the specialist mathematics teachers' approach was directed by their agenda of making the mathematics more visible. Kelan's approach was to explain the mathematics and then reinforce the concept using different representations. Danny's approach was to quickly extract the mathematics from the context so as to reframe the tasks as mathematics, without getting bogged down by the details of the context. Danny introduced the necessary contextual information for a particular contextual rule in a step-by-step manner, thus directing learners to the appropriate point where the mathematics operations appeared as part of the contextual rule. Danny's agenda was also very clear – he wanted his learners to do well in the examinations so he chose to work with past examination papers in class. These specialist mathematics teachers need opportunities to interrogate their own ideas about developing ML skills as compared to learning more mathematics, which is also a need identified by Bolstad (2021) in a study with Norwegian teachers. Bolstad (2021) found that the teachers did not have a clear understanding of what mathematical literacy is and how it is developed. Helping teachers in their learning, is a central issue in research about teacher professional development (Avalos, 2011). Ernest (1989) emphasises that a shift in practice depends fundamentally on the teachers' belief systems, their conception of the nature of the subject, and their mental models of teaching and learning. The findings suggest that ML teachers would benefit from professional development opportunities which can help them understand the purpose of ML. However, researchers have pointed out that there are limited professional development opportunities for ML teachers in South Africa, so this needs to be prioritised by the education department to help teachers better understand the purpose and value of ML (Webb et al., 2013).

In the South African scenario, a further dimension is that the learners who opt to study ML are typically those learners who were not so successful in mathematics in earlier grades. Hence it is incumbent upon ML teachers to try fresh approaches, perhaps those which foreground the context rather than prioritizing the mathematics. Research also suggests that a more learner–centred problem-based approach may prove to be more effective (Hattie & Kimperley, 2007).

The analysis also revealed that the two specialist teachers found it difficult to use contexts to make sense of the problem. The vignettes showed that while Kelan moved flexibly in pointing out commonalities in mathematics concepts across different representations, he found it harder to make links across concepts embedded within contexts. For example, he missed the opportunity to make connections between the concepts of direct and indirect proportion, when they appeared as part of different contexts. Furthermore, he struggled to identify a suitable metaphor to explain the mathematics concept of slope. His attempt to provide access to the notion of slope by using the metaphor of a steep road was misleading since the former implied that a steeper slope was associated with a greater speed while the latter implied that a person would go slower up a steeper slope. Danny too, when working with the conversion of units, opted to justify the conversion by talking about the movement of a decimal comma instead of appealing to the context to make it more meaningful. The two teachers did not consider it important to foreground the context, yet research suggests that when a task has a well-defined context then it can make the mathematics more meaningful, by engaging students and enabling them to attain some personal meaning (Boaler, 1993). This study illustrates that it is not so easy for teachers to recognise authentic situations that can serve as a basis for promoting students' understanding (Turner, 2021). Goos et al. (2014) showed that teachers' understanding of numeracy improved when they worked with researchers in designing tasks for numeracy. Hence this focus could be adopted for future professional development for ML teachers.

It was also found that Edith and Grace, the non-specialist teachers, experienced difficulties in mediating mathematics concepts. Grace had an inadequate understanding of the connections and distinctions between linear relationships, directly proportional relationships and indirectly proportional relationships. She incorrectly believed that any linear relationship that was not a direct proportion would be an indirect proportion. She also faltered in her explanations about discrete and continuous values, showing that she did not quite understand what these quantities were. Edith focused on giving learners rules to solve different problems. In the case of the missing value problem, the solution was reduced to a cross multiplication operation without any explanation of the conditions under which the rule could be applied and the meaning behind the "cross multiplication". The omission of an explanation about why a rule is valid leads to the situation that students may believe that there is nothing special about the conditions and that the rule is also true under other conditions (Wu, 2005). If learners are given rules only without reasons, then mathematics becomes a disconnected collection of rules, making it harder for learners to gain a conceptual understanding.

It was not only with the mathematics that Edith and Grace experienced challenges; they also struggled with the context on many levels. Concerning the context of the freezer, Grace incorrectly explained that the temperature decreased when it was switched off. Grace also misinterpreted the contextual language "hourly fees in excess of eight hours". Edith too, experienced problems with the contextual language. During Edith's discussion of a rate problem, she sidestepped the explanations of the phrase "fuel consumption". Edith's lack of confidence in her understanding of the term meant that she did not spend time unpacking the meaning of the contextual phrase and focused on something else. By providing irrelevant, incorrect descriptions and justifications for contextual elements teachers draw the attention of the learners away from these contextual tools instead of helping them increase their participation within these contexts. Edith offered incomplete explanations for the meaning of the contextual term "average rate", peppered with irrelevant details. Grace too, had problems with negotiating the contextual rules as shown by her incorrect application of the rule for calculating the electrician's fee. Grace's misconception is similar to one associated with the contextual rule for calculating transfer fees (Bansilal et al., 2012). Teachers in the study misinterpreted the transfer duty rule: "For a purchase price of R500,001 to R1,000,000, the transfer duty is 5% on the value above R500,000" and interpreted it as 5% of the amount when it was above R500,000 instead of first working out the difference between the higher amount and R500,000, and finding 5% of that difference. Research in South Africa has identified many challenges faced by students and teachers when working with contextualized tasks, which highlights that working with contextualized tasks can be quite complex (Bansilal et al., 2012; Bansilal, 2017). This is an area that can be targeted by professional development programmes so that ML teachers can find effective ways of mediating such tasks set within contexts.

The teachers showed an over-reliance on, and in some cases a misuse of the textbook. Edith used sets of exercises on ratio and rate problems drawn from a textbook for an earlier grade. Grace followed the textbook together with the teacher's guide faithfully and the sequence was dictated by the text. She struggled with explaining concepts and gave an incorrect answer when the explanation did not appear in the teacher's guide. Studies show that designing tasks and sequencing lessons is a skill that is difficult to master and teachers struggle with these teaching tasks (Goos et al., 2014; Webb et al., 2013). This may explain why these teachers show an over-reliance on the textbook. However, Grace's experiences show that one cannot rely on the textbook only for one's lessons.

7 Conclusion

The curriculum is clear that the aim of ML is not to learn more advanced mathematics but is about using mathematics to make informed decisions in real life settings, and requires the development of skills that are sometimes different from learning mathematics. However, in trying to achieve the mandate of the subject ML, the curriculum and policy intentions can only serve as a guide. Ultimately it is the teachers' beliefs and conceptions about the nature of the subject that influences their practice and approach to the teaching of the subject. In this study we looked at the practices of ML teachers as they mediated ML tasks and concepts in their classrooms. These four cases have helped us identify some of the unique problems that ML teachers can face which can be different from that encountered in teaching mathematics. The results suggest that ML teachers would benefit from professional conversations centred around the purpose and value of ML. There is also a need for professional development opportunities where issues such as the integration of context and content could be interrogated. Some teachers also need opportunities for them to engage more deeply with the mathematics. Researchers have pointed out that there are limited professional development opportunities for ML teachers, which needs to be prioritised by the education department so that teachers' understanding of the role and value of ML is enhanced (Webb et al., 2013). It was found that some teachers rely on the textbook as their guide for teaching. Hence it may be useful for teachers if textbook designers delve deeper into discussions about the contexts and unpack some of the complexities of the contextual language and contextual rules embedded in the various contexts. This will be of benefit to those teachers who struggle with interpreting the contexts themselves.

References

Abdulhamid, L. (2022). Responsive teaching: Differences in primary mathematics teachers' state of awareness and reflections on own teaching. *African Journal of Research in Mathematics, Science and Technology Education*, *26*(2), 89–99. https://doi.org/10.1080/18117295.2022.2092690

Almuna, F. (2016). Investigating the impact of context on students' performance. In B. White, M. Chinnappan, & S. Trenholm (Eds.), *Proceedings of the 39th annual conference of the Mathematics Education Research Group of Australasia* (pp. 102–109). MERGA. https://archive.merga.net.au/Public/Public/Publications/Annual_Conference_Proceedings/2016_MERGA_Conference_Proceedings.aspx

Avalos, B. (2011). Teacher professional development in teaching and teacher education over ten years. *Teaching and Teacher Education, 27*, 10–20. https://doi.org/10.1016/j.tate.2010.08.007

Bansilal, S. (2017). The application of the percentage change calculation in the context of inflation in mathematical literacy. *Pythagoras, 38*(1). https://doi.org/10.4102/pythagoras.v38i1.314

Bansilal, S., & Debba, R. (2012). Exploring the role of contextual attributes in a Mathematical Literacy assessment task. *African Journal for Research in Science, Mathematics and Technology Education, 16*(3), 302–316. https://doi.org/10.1080/10288457.2012.10740747

Bansilal, S., Mkhwanazi, T. W., & Mahlabela, P. (2012). Mathematical literacy teachers' engagement with contextual tasks based on personal finance. *Perspectives in Education, 31*(3), 98–109. https://journals.ufs.ac.za/index.php/pie/article/view/1774

Bansilal, S., & Ubah, I. (2020). The use of cross multiplication and other mal-rules in fraction operations by pre-service teachers. *The Journal of Mathematical Behavior, 58*. https://doi.org/10.1016/j.jmathb.2020.100781

Boaler, J. (1993). The role of contexts in the mathematics classroom: Do they make mathematics more "real"? *For the Learning of Mathematics, 13*(2), 12–17.

Bolstad, O. H. (2021). *Mathematical literacy in Norway*. Paper presented at the ICME 14, Shanghai, China.

Department of Basic Education (DBE). (2011). *Curriculum and assessment policy statement: Further education and training phase Grades 10–12*. Mathematical Literacy. Government.

Department of Basic Education (DBE). (2012). *November examination 2012*. Grade 11 Mathematical Literacy – Paper 2. Government.

Department of Basic Education (DBE). (2015). *Mind the gap*. Mathematical Literacy study guide Grades 10–12. Government.

Duranti, A., & Goodwin, C. (1992). *Rethinking contexts: Language as an interactive phenomenon.* Cambridge University Press.

Ernest, P. (1989). The knowledge, beliefs and attitudes of the mathematics teacher: A model. *Journal of Education for Teaching: International research and pedagogy, 15*(1), 13–33. http://dx.doi.org/10.1080/0260747890150102

Gal, I., Grotlüschen, A., Tout, D., & Kaiser, G. (2020). Numeracy, adult education, and vulnerable adults: A critical view of a neglected field. *ZDM–Mathematics Education, 52*(3), 77–394. https://doi.org/10.1007/s11858-020-01155-9

Geiger, V. (2021). *How teachers generate ideas for classroom numeracy tasks.* Paper presented at the ICME 14, Shanghai, China.

Graven, M., & Venkat, H. (2007). Emerging pedagogic agendas in the teaching of mathematical literacy. *African Journal of Research in Mathematics, Science and Technology Education, 11*(2), 67–84. https://doi.org/10.1080/10288457.2007.10740622

Greeno, J. G. (1991). Number sense as situated knowing in a conceptual domain. *Journal for Research in Mathematics Education, 22*(3), 170–218.

Goos, M., Geiger, V., & Dole, S. (2014). Transforming professional practice in numeracy teaching. In Y. Li, E. Silver, & S. Li (Eds.), *Transforming mathematics instruction: Multiple approaches and practices* (pp. 81–102). Springer.

Hattie, J., & Timperley, H. (2007). The power of feedback. *Review of Educational Research, 77*(1), 81–112. https://journals.sagepub.com/doi/abs/10.3102/003465430298487

Hoogland, K. (2021). *Common European numeracy framework – A multifaceted perspective on numeracy.* Paper presented at the ICME 14, Shanghai, China.

Mahlabela, P. T., & Bansilal, S. (2015). An exploration of learners' theorems-in-action used in problems on ratio and proportion. *Pythagoras, 36*(2), 11–20. http://dx.doi.org/10.4102/pythagoras.v36i2.252

North, M., & Christiansen, I. M. (2015). Problematising current forms of legitimised participation in the examination papers for mathematical literacy. *Pythagoras, 36*(1). https://doi.org/10.4102/pythagoras.v36i1.285

OECD. (2018). *PISA 2022 Mathematics framework (draft).* https://pisa2022-maths.oecd.org/files/PISA%202022%20Mathematics%20Framework%20Draft.pdf

Pillay, E. (2018). *Exploring mathematical literacy teachers' classroom practices.* [Unpublished doctoral thesis, University of KwaZulu-Natal]. Durban.

Polkinghorne, D. E. (1995). Narrative configuration in qualitative analysis. In J. A. Hatch & R. Wisniewski (Eds.), *Life history as narrative* (pp. 5–23). Falmer.

Savard, A., & Cavalcante, A. (2021). Financial education and mathematics education: A cross-cutting analysis of the epistemological intersection of financial numeracy. In A. Savard & A. Cavalcante (Eds.), *Financial numeracy in mathematics education: Mathematics education in the digital era* (Vol. 15, pp. 3–8). Springer. https://doi.org/10.1007/978-3-030-73588-3_1

Sawatzki, C., Brown, J., & Powers, T. (2024). Supporting teachers to develop mathematical literacy through contemporary financial contexts. In O. H. Bolstad, S. Goodchild, & M. Goos (Eds.), *International perspectives on teaching and learning for mathematical literacy* (pp. 125–151). Brill.

Turner, R. (2021). *Mathematical literacy – What, why and how.* Paper presented at the ICME 14, Shanghai, China.

Venkat, H., & Winter, M. (2015). Exploring the nature and coherence of mathematical work in South African mathematical literacy classrooms. *Research in Mathematics Education, 12*(1), 53–68. https://doi.org/10.1080/14794800903569865

Webb, L., Bansilal, S., James, A., Goba,B., & Khuzwayo, H. (2013). Curricular reform for the teaching of Mathematical Literacy teachers in SA: Looking back to move forward. In L. Wood & C. C. Wolhuter (Eds.), *Criticality, creativity and connections: In pursuit of educational innovations.* Proceedings of the 2012 Annual Conference of EASA (pp. 431–443). EASA.

Wu, H. (2005). *Must content dictate pedagogy in mathematics education?* Talk given at California State University, USA.

PART 2

Designing and Evaluating Mathematical Literacy Tasks

CHAPTER 4

Levels of Context Use

An Analytical Tool to Inform the Influence of Problem Context on Students' Performance and to Construct Mathematical Problems

Felipe Almuna and Kaye Stacey

Abstract

One reason why mathematics is a compulsory educational subject globally is the fact that mathematical knowledge is applied in multiple and different areas outside mathematics. The PISA mathematics frameworks see mathematics as an essential tool for students beyond school as they face issues and challenges in the diverse contexts where mathematical knowledge is relevant. Despite many mathematics curricula around the globe recommending teachers use real-world contexts in teaching, data from PISA show consistently that a high percentage of students do not reach the minimum standard of mathematical literacy that enables them to deal with real-world contexts. This finding relates to a long-standing issue of the educational research on mathematical modelling to investigate how problem context influences students' performance. In this vein, this chapter (i) offers a definition of the context of a mathematics problem, (ii) presents a concise classification of the "level of context use" required to solve a mathematical problem, (iii) briefly reports the results of a study that shows that a higher level of context use makes problems more difficult and also gives some insight into why this happens, and (iv) discusses issues of problem development that face teachers and others when they wish to assist students to work more effectively with problems in context.

Keywords

problem context – real world problem solving – levels of context use – mathematics performance – construction of mathematical problems

 | DOI:10.1163/9789004710962_004

1 Introduction

The incorporation of context in mathematics problems has been highly recommended by current reform documents and mathematics curricula around the globe (e.g., Australian Curriculum Assessment and Reporting Authority, 2016; Finnish National Agency for Education, 2021; Ministry of Education Chile, 2016). These documents recommend that students learn to mathematise problems stated in a realistic or real-world context (i.e., to transform them to a strictly mathematical form) not just by applying procedural and conceptual mathematical knowledge, but also by interacting with the problem context. Relationships between problems arising in real-world contexts and Mathematics are important in tests in many countries but particularly relevant to the Programme for International Student Assessment (PISA). PISA mathematics does not assess any particular curriculum but is designed to assess mathematical literacy, a students' capacity to apply their knowledge to real-world contexts. As observed by Blum (2002, p. 151) the importance of the PISA definition of mathematical literacy is that it involves an interpretation and understanding of mathematics put to functional use in a multitude of different situations. Therefore, the assessment of mathematical literacy must include items that involve thinking about the context in which the problems are embedded.

PISA results demonstrate that some countries, like Chile, have a high percentage of students not reaching the minimum standard at which they begin to demonstrate the mathematical literacy competencies to deal with real-world contexts (see OECD, 2019). Although the difficulties secondary students experience in applying mathematics to problems in context are particularly relevant to PISA, it is also a long-standing issue in mathematics education research (Niss et al., 2007; Stacey & Turner, 2015). The long tradition of educational research on word problems and on mathematical modelling has scrutinised students' mathematical interactions with real-world context from the shortest exercises to modelling problems that are much more substantial than the PISA items. Nevertheless, there is much more to discover about how a real-world problem context may influence the students' performance.

Therefore, after a short review of relevant constructs from PISA, the central ideas of this chapter are to (i) define the context of a mathematics problem (ii) illustrate the "level of context use" demanded by a mathematical problem and present a refined classification of this construct, (iii) briefly report some empirical research on how the level of context use affects students' performance and (iv) demonstrate how level of context use can be used as an analytical tool when constructing mathematical problems for the mathematics classroom considering issues that teachers and problem designers face when creating problems embedded in real-world contexts. In this vein, this chapter capitalises on the interest in PISA around the world by giving a further insight into how "problem context" may influence students' performance. Therefore, it has implications for using problems in context in the mathematics classroom. As this chapter has a focus on some PISA constructs, it begins by offering a concise introduction to key aspects of PISA that are directly relevant to this chapter.

2 Key Aspects of PISA Mathematics

For the Organisation for Economic Co-operation and Development (OECD), mathematics is seen as an essential tool for students as after their compulsory education, they will face challenges in which mathematical knowledge is relevant, in their personal and working lives and in their roles as informed citizens. Hence, educational systems should prepare students for this. Therefore, the 2018[1] PISA theoretical framework for mathematics offers mathematical literacy as the key construct that PISA surveys assess and defines it as:

> An individual's capacity to formulate, employ and interpret mathematics in a variety of contexts. It includes reasoning mathematically and using mathematical concepts, procedures, facts and tools to describe, explain and predict phenomena. It assists individuals to recognise the role that mathematics plays in the world and to make the well-founded judgements and decisions needed by constructive, engaged and reflective citizens. (OECD, 2019, p. 75)[2]

As is clear in the definition, the construct of mathematical literacy strongly emphasises the need to develop students' capacity to use mathematics in context. PISA analytical frameworks from 1999 to 2018 included a set of fundamental mathematical capabilities (also labelled mathematical competencies) that underpin the processes needed by students to use their mathematical knowledge and skills. They are used to build the proficiency descriptions through which the OECD reports students' levels of mathematical literacy to (i) formulate situations mathematically, (ii) employ mathematical concepts, facts, procedures, and reasoning, and (iii) interpret, apply and evaluate mathematical outcomes (OECD, 2019). Briefly and in an idealised scheme, formulation occurs when the student moves from a problem expressed in everyday language to a problem described in abstract mathematical terms. Students then employ mathematical facts and skills (computations, algebraic manipulations, geometric deductions, construction of statistical charts, etc.) to arrive at a mathematical solution. Then students need to interpret their mathematical solution in real world terms and evaluate its appropriateness as a real-world

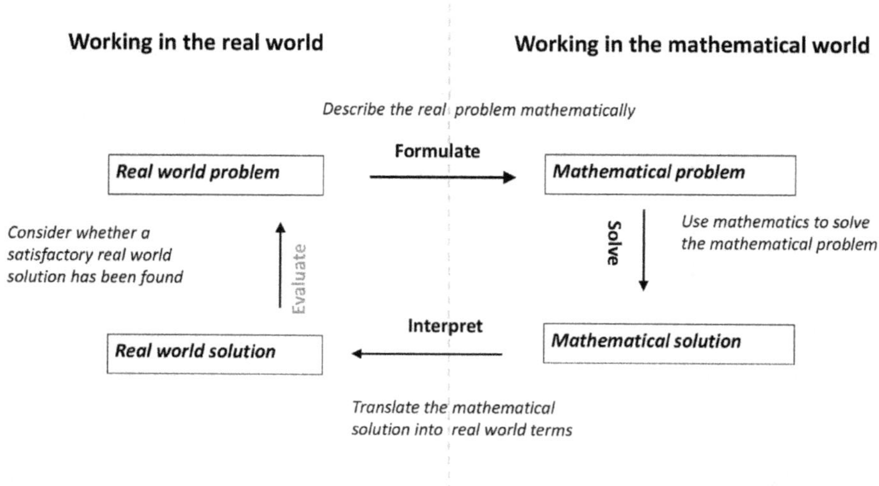

FIGURE 4.1 The PISA modelling cycle

solution. Both capabilities and processes are part of student's understanding of mathematical ideas and capacity to apply his or her mathematical knowledge in the real world (Turner & Adams, 2012). The various PISA frameworks have always drawn significantly on the conceptualisation of mathematical modelling (Stacey, 2015) that is captured in the image of the modelling cycle of Figure 4.1. Researchers in mathematical modelling frequently refer to such a cycle, very frequently substantially elaborated from the simple diagram below used by PISA for the 2012, 2015 and 2018 surveys. In addition to noting that the modelling cycle is an idealised description, note that individual items in a large-scale assessment such as PISA can only involve students in some stages of the modelling cycle and cannot realistically expect students to engage in any thorough evaluation. Interpretation and evaluation (see Figure 4.1) are therefore reported together. PISA items represent one end of the spectrum of mathematical modelling problems that have been considered in the mathematics education literature (Almuna, 2020). Of special interest to this chapter are the two horizontal arrows (formulating situations mathematically and interpreting, applying and evaluating mathematical outcomes), which represent the movement from the mathematical world to the real world and back. These are the two processes that link the real-world context and the mathematics.

3 Defining the Context of a Mathematics Problem

There are various definitions in the literature for the word "context" when it is used to describe the real (or fantasy) world setting of a mathematics problem.

Note that the word "context" is also frequently used with a very different meaning, describing the social, physical, and other conditions under which mathematics is done. For example, in this other sense, problems might be done in the "context" of a multilingual classroom familiar to all students, as part of a timed test, with calculators allowed, late in the afternoon. Other terms used to describe problem context in our sense are situation (OECD, 2019; Stacey, 2015, p. 74), setting, or cover story (often used in relation to word problems; e.g., see Silver, 1981), and thematic content (often used in reasoning studies; e.g., see Ross et al., 1986). Clear definitions of context, therefore, are difficult to find in the literature but classifications of context and the roles that they play in teaching and problem solving abound. For further discussion of classifications, roles and relevance of mathematical problem contexts, refer to Crisp and Greatorex (2022), Brown (2019), Greatorex (2014), and Almuna and Stacey (2014).

For this chapter, the definition of the "context" of a mathematics problem follows the first author's doctoral thesis:

> [Context is] all the information about the real-world situation that is contained in and can be derived using common knowledge from the statement of a mathematical problem. The contained and derived information might be necessary or unnecessary for the mathematisation and solution of the problem. The context is distinct from the problem's stimulus but accessed through it. (Almuna, 2020, p. 20)

To further clarify the definition of context and other terminology used in this chapter, consider the problem in Figure 4.2, which is a modification of a released PISA problem Rock Concert M552Q01 (OECD, 2006). Figure 4.2 shows the stimulus, the actual material (i.e., a written mathematical problem) that is presented to students. The stimulus of a problem includes the words and their physical layout and font, pictures, graphs, diagrams, given formula, and any multimedia material. The stimulus is distinguished from content-related features of the wording such as grammatical structure and readability.

For a rock concert, a rectangular field of size 50 m by 25 m was reserved for the audience. The concert was completely sold out and the field was full, with all the fans standing. Which one of the following is likely to be the best estimate of the total number of people attending the concert?

A) About 1 250 B) About 2 500 C) About 3 750 D) About 5 000

FIGURE 4.2 Modified Rock Concert PISA problem[3]

The context of this problem involves a concert, rock music and a field with fans standing. Understanding the context involves aspects such as appreciating the general size and shape of the field, the nature of a rock concert, likely density of the crowd at a sold-out concert, that facilities for the crowd are probably outside this area, and more general aspects of any concert including the purchasing of the tickets, venue details (e.g., in the centre of a stadium or in a country paddock), whether there are emergency exits or aisles. However, not everything about the context is necessary to mathematise the problem. In fact, only the dimensions of the field (given) and some way of quantifying the likely density of the crowd at a rock concert are needed for the calculation. The likely density of the crowd is not supplied but is assumed by the problem creators to be able to be estimated adequately by nearly all target students relying on their common knowledge of this real-world context (this assumption may be false, perhaps one of the reasons why M552Q01 was used only in the field trial and never in a main PISA survey).

The term "mathematical core" refers to the abstract mathematics underlying the solution including the set of variables, assumptions and relationships required for the solution, so it includes mathematical concepts, the model, calculations, and mathematical strategies. The mathematical core of this item, the estimation of a static crowd, is straightforward (i.e., area of the field occupied by fans multiplied by the estimated likely average density of the crowd). However, whilst there are widely agreed numerical densities for crowds (e.g., four persons per square metre in a dense crowd according to Watson & Yip, 2011, p. 105), it was expected that students would not know these but would make their own estimates of typical body dimensions and spacing at such a concert to solve this problem. Obviously, there are other methods of estimation, for instance, students can estimate density simply by visualising how many people are in a crowded square metre, or they might imagine fans lined up in rows and columns along the full length and width of the field, with say 30 cm between fans. These are variations on the basic mathematical core of the problem. With any mathematical method, the context clues that the field was completely sold out and fans were standing are needed to guide students in their estimation. The multiple-choice format further helps as it implicitly indicates the required accuracy of the estimation and gives students a range of reasonable answers (e.g., option A at 1 person per square metre is a loose crowd). For this item, students not only have to make assumptions from their real-world knowledge to formulate the problem mathematically but also need to interpret the solution against the context: whether it has produced a reasonable estimate of the number of people for this kind of concert (e.g., "I assumed that 4 people can be accommodated per square metre and that gives a number of people that seems to be a reasonable estimate for a rock concert"). In their interpretation, some students may also have to consider whether their mathematical answer is better rounded up or down to match one of the multiple choices provided. It is interesting to consider how living with COVID-19 may have impacted on students' approach to the original rock concert item. In a world with COVID, many students may want to incorporate social distancing in their estimates and may even know recommended spacing in crowds. The multiple-choice options would have to change.

4 Levels of Context Use

Among the many classifications of contexts, the "levels of context use" introduced by Jan de Lange is the focus of this chapter. de Lange (1979, 1999) classified problems according to the degree to which solving the problem requires students to engage with the problem context. The classification is not well understood yet, hence our attention. In his 1979 scheme, de Lange distinguished four "levels of context use" (zero, first, second and third levels). The first three of these levels are discussed below. They are relevant to the relatively short, written problems common in many classrooms, including PISA items, and hence to this chapter. de Lange's final "third" level is not used here. It relates to contexts that allow students to discover, construct or reinvent new mathematical concepts. These contexts have a special place in Realistic Mathematics Education (RME) where they provide an alternative scenario for students to "learn mathematical concepts and structures that are relevant for the problem situation" (van den Heuvel-Panhuizen, 2005, p. 13) and are used as vehicles to support growth of understanding of major mathematical topics (Treffers, 1987).

A good description of the levels of context use in PISA is given in the expert commentary "Learning Mathematics for Life" (OECD, 2009a). This section summarises this description, but we also point out ways in which the description is narrow or vague. It is important to note that PISA has never used levels of context use as a reporting category or allocated any fixed percent of items to each level. Rather the intention has been that some items at the higher levels of context use would be included.

Problems with zero-order level of context use are described as those where context "is used to make the problem look like a real-world problem" (OECD, 2009a, p. 30). In the literature this is sometimes called fake context or camouflage context or Stillman's (1998) context as a border around a problem (rather than a wrapper or a tapestry). After a comment that such problems are not common amongst PISA problems, the expert commentary gives as an example the item M266Q01 Carpenter (see Figure 4.3).

Question 1: CARPENTER *M266Q01*

A carpenter has 32 metres of timber and wants to make a border around a garden bed. He is considering the following designs for the garden bed.

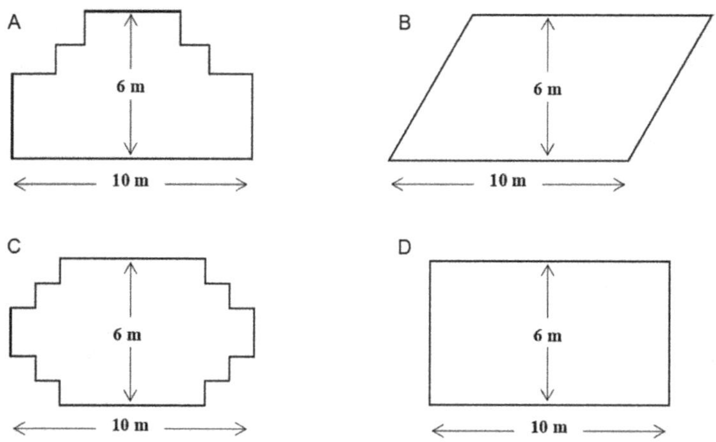

Circle either "Yes" or "No" for each design to indicate whether the garden bed can be made with 32 metres of timber.

Garden bed design	Using this design, can the garden bed be made with 32 metres of timber?
Design A	Yes / No
Design B	Yes / No
Design C	Yes / No
Design D	Yes / No

FIGURE 4.3 Item M266Q01 Carpenter (OECD, 2006)[4]

This problem gives four possible designs for a garden and students have to identify those designs for which a border can be made with 32 metres of timber. The expert commentary observes that nothing about the carpenter or the garden is needed; the context only serves to camouflage the mathematics of finding the perimeter of the given shapes. Students have to know what a carpenter and a border are, but there seems to be no other understanding of the context required. Although not included in the OECD (2009a) description, it seems to us that many genuine but routine mathematics problems should also come under zero-order level of context use. For example, if the price of a $145 long distance train ticket is to be increased by 5%, students can apply routine knowledge to find the new price. They do not need to know anything specific about trains or ticket purchase or ticket pricing. We also expect that intra-mathematical problems with no real-world context should be classified as zero-order.

Next comes first-order level of context use. This use takes place "when the context is relevant and needed for solving the problem and judging the answer" (OECD, 2009a, p. 31). The example given is item M465Q01 Water Tank (OECD, 2006), shown in Figure 4.4.

A water tank comprising a conical part at the bottom and a cylindrical part above is shown, with dimensions. Students are told the tank begins empty, then

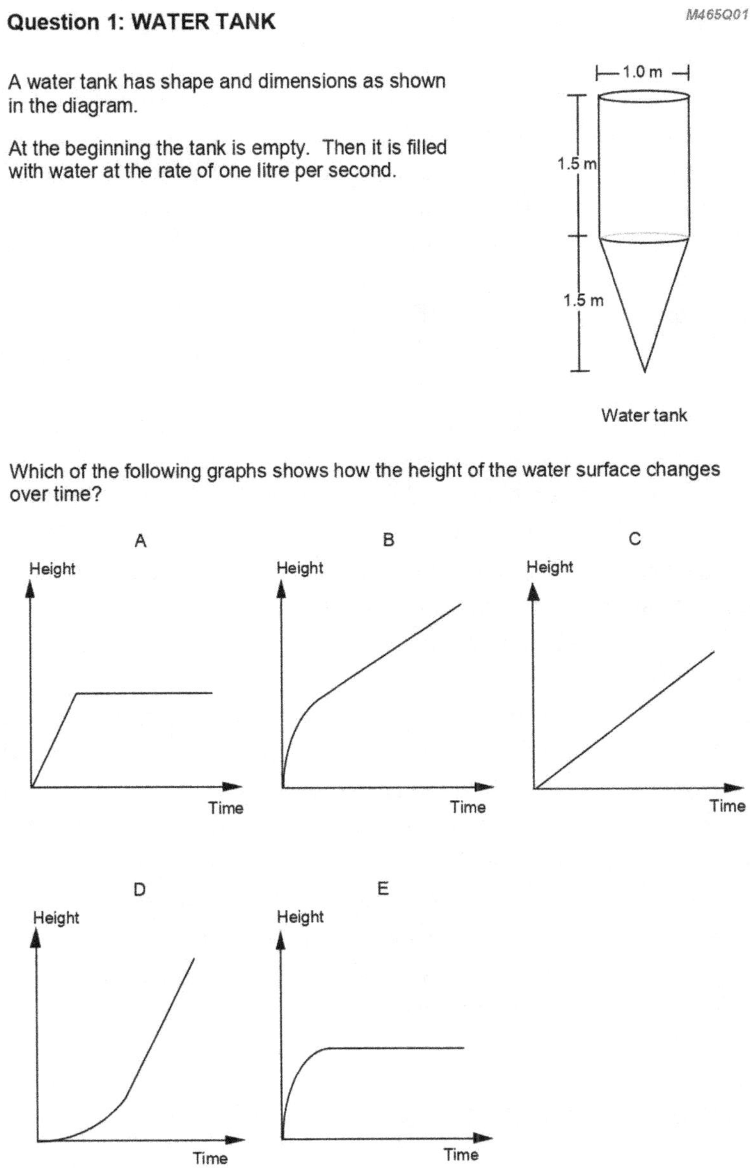

FIGURE 4.4 Item M465Q01 Water Tank (OECD, 2006)[5]

it is filled with water at the rate of one litre per second. Five possible sketch graphs of height of the water surface against time are shown. Students have to choose the most appropriate one. To solve this problem, students have to think about how the steady increase of volume of water changes the height of the water surface and hence also the surface area, and hence the rate at which the surface rises. Context facts such as the cone being filled before the cylinder, that it is a fluid being poured in and the qualitative relationship between surface area and change in volume, are needed for the solution, creating first-order level of context use. The Carpenter problem (Figure 4.3) could become a first-order level of context use problem if the length of timber needed was not simply the geometric perimeter of the shapes. For example, overlap at corners may be needed to make joints (extra length dependent on timber thickness) or there may be wastage from cutting from individual planks. In such variations, the context of a carpenter with 32 metres of timber to make a border around a garden bed would not act only as a camouflage.

Next, the second-order level of context use requires a strong interaction between the student and the context. "[It] appears when one really needs to move backwards and forwards between the mathematical problem and its context in order to solve the problem or to reflect on the answer within the context to judge the correctness of the answer" (OECD, 2009a, p. 31). This level of context use is illustrated with the modified Rock Concert PISA problem, as discussed above. To solve this problem, students must bring their own estimate of the area that a person would take up in such a concert into the calculations. That information (or an equivalent) is missing from the problem statement, and it must be drawn from knowledge of the context in formulating the problem mathematically.

When we tried to classify problems according to their level of context use, we found that the above descriptions need clarification, refinement and further operationalisation and validation. For example, the OECD (2009a) descriptions blur first-order and second-order level of context use. What, for instance, is the intended meaning of, and difference between, context that is "needed" (first-order) and "really needed" (second-order)? What actually signifies a "move backwards and forwards between the mathematical problem and its context"? As is discussed below, a new set of rules for categorising problems according to level of context use is required.

5 Does Level of Context Use Influence Student Performance?

It is agreed in the literature that there is value in assessing students' capacity to solve mathematical problems with second-order level of context use, because such problems require students to deal to some degree with context features in the same way as they arise when using mathematics to solve problems outside of school (OECD, 2015, 2019; Stacey, 2015). This is because, as the level of context use of a problem increases, so "does the importance of being able to imagine aspects of the real-world and the potential role of mathematics within it" (Sawatzki, 2017, p. 28). In this vein, the OECD (2009a) also notes: "The highest levels of mathematical literacy involve the ability to effectively handle such second-order contexts. This is the essence of real-world problem solving" (p. 31).

5.1 *Preliminary Investigation*

Although there are many studies on the effect of other aspects of context on student performance (e.g., context familiarity), level of context use is a classification with very few empirical studies. Almuna used levels of context use as part of the set of analytical tools to investigate the effect of an alteration of context on students' performance (Almuna, 2010, 2016; Almuna & Stacey, 2014). His mixed-method study examined differences in students' performance when they solved both PISA problems and matched problems that were designed with the same mathematical core but with a different context. Findings indicated that a change of context influenced students' performance more for problems that involved the second-order level of context use than zero or first-order levels of context use. The data set (using just two pairs of items for each level of context use) was too small to make a strong claim, but the result makes sense. With second-order level of context use problems, the information brought in from the context by the student plays a role in the solution, making success more dependent on the actual context chosen. Data from interviews revealed some of the ways in which students did or did not use their context knowledge, and how it sometimes assisted and sometimes blocked a good solution. These results are discussed below. This preliminary study revealed that differences between student performance related to level of context use are worth studying, and also that a refined set of classification rules to allocate problems to level of context use was required.

5.2 *Refined Classification System for Level of Context Use of Problems*

The refined classification introduced here builds on previous theory (de Lange, 1979, 1999; OECD, 2009a) to create a framework that allows more reliable classifications of the levels of context use. Hence, the starting point for the refinement was to re-define the construct of levels of context use, which was understood as:

> The degree to which solving a problem requires engagement with the context to formulate a problem in mathematical terms, solve it in its mathematical form, and to interpret and validate the answer in relation to the context of the given problem. (Almuna, 2020, p. 25)

In terms of the PISA modelling cycle (Figure 4.1) engagement with the context principally occurs during the two processes of (i) formulating situations mathematically, and (iii) interpreting, applying and evaluating mathematical outcomes. In the new scheme, the nature of use of context is separately scored for each of these two processes, and the scores are added to decide the levels of context use. Figure 4.5 illustrates the scheme. There is a set of indicators for each process. For the classification, heuristic ordinal scores 0, 1, 2 are assigned to the indicators. The values simply indicate that there are differences among indicators in terms of the level of context use. For instance, assigning a code of A0 to a problem indicates that the "use of context for formulation" is, at least for the majority of students, likely to be at a minimum level; whereas assigning B2 indicates that consideration of the context when interpreting and evaluating the mathematical result in real world terms is at a maximum level, judged relative to the set of problems under consideration, in the present case PISA and PISA-like items, with the PISA target student population. The level of context use for a problem is then decided by the sum of the two scores, as is shown at the bottom of Figure 4.5.

The scale was not statistically created; it rather used a conceptual definition of levels of context use built on judgements of how context was used in the formulation and interpretation processes of the PISA modelling cycle. The combination of these criteria and indicators reflects conceptually broadly increasing levels of context use but does not reflect problem difficulty per se. Problem difficulty depends on many other factors. Lastly, details of the rationale, construction, combinations of codes and a multiple-rater validation of the scheme and the choices of cut-off points can be found in Almuna (2020).

The formulation criterion, that is, the formulation of the mathematical problem, refers to the different degrees to which the problem context influences putting mathematics into practice. In general, context may offer room to put mathematics into practice straightforwardly or may require devising a tailored mathematical model. The latter situations require the solver to interact with the context; this interaction with the context may happen at different levels while formulating the problem. To clarify this explanation, in problems classified with the A0 code, direct mathematical actions can be taken from instructions given, without any significant consideration of the context. For example, the Carpenter problem (Figure 4.3) is a geometry problem, requiring

Codes, Description of Indicators and Scores		
Formulation of the mathematical problem		
Code	***Description of Indicators***	***Score***
A0	Context is not needed for formulating the problem in mathematical terms. From instructions given, direct mathematical actions can be taken.	0
A1	Formulating the problem mathematically requires more than just straightforward actions to be taken from the given instructions. One has to identify and represent mathematically relevant information, variables, relationships, or assumptions. Context provides cues for this.	1
A2	Formulating the problem mathematically requires further consideration of the context to retrieve relevant information, variables, relationships, or make assumptions, which are not explicitly provided.	2
Interpretation of mathematical results		
Code	***Description of Indicators***	***Score***
B0	Context is not needed to interpret, decide, or evaluate mathematical results or arguments.	0
B1	Context is used to decide whether the mathematical results or a mathematical argument satisfy the requirements of the problem.	1
B2	Context is used to judge the adequacy of the mathematical results and arguments in terms of the assumptions made for the mathematical formulation of a problem. That is to say, context and mathematical results or arguments need to be reconciled globally for a valid solution that satisfies the requirements of the problem.	2
Combining the scores		

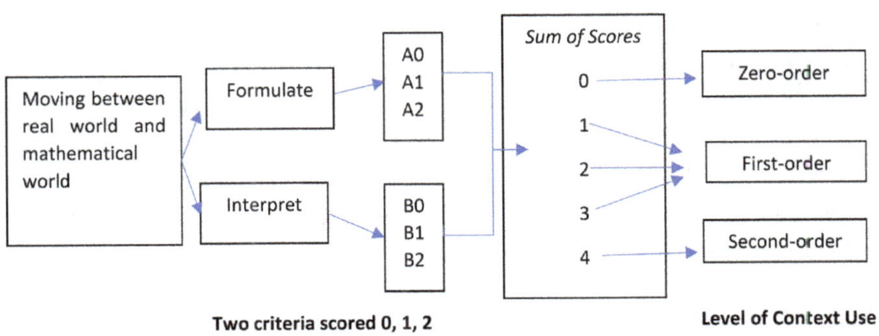

FIGURE 4.5 Classifying problems according to levels of context use, by combining scores on two processes

knowledge of various properties of lengths of lines but no knowledge of making garden borders, so is coded A0. The Carpenter problem is not easy, but this is not due to the demands of formulation related to the context. Other problems coded A0 are those that require only the application of well-known procedures, often taught to students several years before the PISA target age. In the case of percentages, for instance, students have previously learned to identify a broad class of problems where that mathematical content is commonly used. Other standard models of wide application that students learn to apply include finding totals and differences, equal sharing, comparing proportions, setting up and solving simple equations. In problems coded A0, it is expected that students at the target age can identify the relevant mathematical model (including the salient variables, relationships, and assumptions) straightforwardly.

In problems classified as A1, when formulating problems mathematically the contexts require students to identify and translate information to solve a problem mathematically, identifying relevant variables and relationships between them or making assumptions from clues given in the context. For example, in the Mine Schedule problem (Figure 4.6), relevant data are carefully described in the statement of the problem. The problem can be formulated mathematically based on the mathematical concept of periodicity. Information about the cycles is given (i.e., roster and transport services cycles), but not about the time they took (i.e., flight travelling time). Students need to grasp

Mine Schedule

Martin works as a mining employee in a remote place. Because of this, Martin has to be flown to his work site instead of relocating him and his family permanently. His contract starts on the 1st of October. Then, he works at the mine for a continuous period of 15 days. After that he is flown back home and in 20 days he is back at the mine again. This process is repeated regularly over Martin's contract period. Martin wants to know where he will be in 4 and half months. Consider 1 month has 30 days.

Show your working and explain your reasoning.

A) At the mine
B) On his way home
C) At home
D) On his way to the mine

Bus Schedule

In a small city, there is only one airport shuttle bus. The first service starts at 7am; then the bus leaves the bus station towards the airport. It arrives at the airport after 15 minutes. 20 minutes later the bus is back to the bus station ready to depart again. This process is repeated regularly until the last service at night. Martin, a backpacker, wants to catch this bus service to the airport. He wants to know where the bus will be at 11:30am. Consider 1 hour has 60 minutes.

Show your working and explain your reasoning.

A) At the bus station
B) On its way to the airport
C) At the airport
D) On its way to the bus station

FIGURE 4.6 Two problems at first-level of context use (Almuna, 2020)

that a full cycle takes 35 days. The mathematisation of the problem is that at t = 35 the cycle began again.

In problems classified as A2, students need to draw on their own knowledge of the context to retrieve extra pieces of information not explicitly stated, selecting important variables, or identifying non-standard relationships to formulate a mathematical representation of the problem and/or data to apply it. This was discussed above for the modified Rock Concert PISA problem (Figure 4.2).

After completing the mathematical formulation of a problem in context, relevant mathematics concepts and tools are employed to solve it mathematically. Subsequently, in the interpretation process the mathematical results may need to be interpreted or validated in terms of the real-world context. Hence, the interpretation criterion identifies differences in context use when linking the results to the problem context. For problems that can be classified into a B0 code, mathematical results are directly usable to answer the real-world problem (e.g., finding a cost after a given percentage increase), or as in the Carpenter problem (Figure 4.3), where interpretation in relation to the context is barely necessary. The other case is when the mathematical results of a problem were needed to be interpreted or justified within the context; that is to say, mathematical results need to satisfy the real-world requirements of the problem. For instance, in the Mine Schedule Problem (Figure 4.6) context provides a chance for students to interpret and translate information mathematically and the judgements need to be interpreted in relation to the context of the problem. The mathematical solution shows that after 135 days, Martin will be 30 days into a cycle. Since his new work cycle does not start for 5 days, he is very unlikely to have started his flight(s) back to work. Making this judgement in the real-world context produces a B1 rating. In contrast, in the modified Rock Concert PISA problem, interpreting (and evaluating) the mathematical answer obtained from a calculation involves a more global assessment of whether a reasonable number of people for a rock concert has been found from the model used and the data assumed (e.g., "I assumed that 4 people can be accommodated per square metre and that every square metre was available for fans to stand"). In their interpretation, students may also have to consider whether, given the context, their mathematical answer is better rounded up or down to select one of the multiple choices provided. This problem is therefore classified into a B2 code.

Taking the above into account, three different levels of relations in the interpretation process were theoretically identified. Sometimes interpretation is not needed, because the context is not needed, or it was disregarded for the mathematical formulation. In other cases, context is used to decide whether the mathematical results or a mathematical argument satisfy the requirements of the problem. Thirdly, context is sometimes used to judge the adequacy of the mathematical results and arguments in terms of the assumptions made for the mathematical formulation of a problem. In those cases, codes B0, B1, and B2 are assigned respectively, and their numerical values are respectively 0, 1, and 2.

Finally, the level of context use of a problem is determined by adding the sub-scores from the A and B criteria. If the total score is 0, the problem is classified as zero-order level of context use. This can only happen if the problem is rated as (A0, B0) meaning there is no significant demand arising from the real-world context on either criterion. If there is high demand on both criteria, the problem is rated (A2, B2) and classified as second-order level of context use. Problems with intermediate combinations of demand on these two criteria (scores of 1, 2 or 3) are allocated to the first-order level of context use. Further justification of the choices of cut-off points can be found in Almuna (2020). As the examples above illustrate, the new scheme is intended to match the original conceptualisation of levels of context use put forward by de Lange and later the OECD expert commentary, whilst accepting that the classification of some problems may change. However, by making the criteria more precise, the new scheme provides a strong basis for future research.

5.3 *The Effect of Level of Context Use on the Performance of Year 10 Students*

The effect of level of context use on students' performance was investigated as part of Almuna's (2020) doctoral study, which also examined the effect of context familiarity and engagement (which are not reported here). One hundred and fifty-one students from eight Year 10 classes at one volunteer school in Melbourne, Australia, solved a total of nineteen problems in a paper-and-pencil format. Because of potential inter-problem effects and limited class time for testing, a rotated design of eight booklets was used. Each student solved three problems, one at each level of context use. Problems used were slightly adapted PISA items and specially constructed PISA-like items. Altogether there were 6 families of problems. The problems in a family differed in context but were all carefully matched so they had the same level of context use and the same mathematical core, drew on the same mathematical competency(ies), and had closely matched structural and stylistic features of the stimuli. For more detailed information, see Almuna (2020). As an example, the modified Rock Concert PISA problem (Figure 4.2) was in Family E with two very similar problems: estimating the number of cars that could travel on a ferry with a 50m by 25m car space and estimating the total number of people who could sleep

TABLE 4.1 Distribution of ratings of 453 solutions from 151 students by levels of context use and statistics from ordinal logistic regression model (with zero-order level of context use as reference)

Level of context use	Type and number of solutions			Odds ratio	95% confidence interval
	No credit	Partial credit	Full credit		
Second-Order	65	52	34	6.64	(2.68, 16.41)
First-Order	28	63	60	2.63	(1.67, 4.12)
Zero-Order	30	14	107	1	
Total number of solutions	123	129	201		

on beds at an emergency evacuation centre. Other families of problems emulated M177Q01 Robberies (OECD, 2006, p. 34) and Seal (OECD, 2009b, p. 117), again with altered contexts. The issue of constructing well-matched PISA-like items using levels of context use will be discussed at the end of this chapter.

To test the relative effect of degrees of levels of context use on students' performance, an ordinal logistic regression analysis (McCullagh & Nelder, 1989) was conducted. Table 4.1 presents the distribution of students' scores across levels of context use, as well as a summary of the ordinal regression containing the odds ratios (both of which are highly statistically significant), and the confidence intervals. The number of solutions receiving full credit shows clearly that students' performance decreased as levels of context use increased.

The regression analysis shows that the odds of a student obtaining full credit versus no-credit performance were nearly seven times higher on zero-order level of context use problems than on second-order level of context use problems and nearly three times higher on zero-order level of context use problems than on first-order level of context use problems. Level of context use influenced performance more than context familiarity or engagement. Although there are technical and methodological aspects that could have impacted in this result – which are discussed in Almuna (2020) – the above results on levels of context use are best explained by the fact that the second-order level of context use problems used in the study required more careful consideration of a problem's real-world context than first- and zero-order use of context problems, and hence more elaboration in their solutions.

What is the likely explanation of this link between levels of context use and item difficulty? A first possibility is that students did not possess any appropriate real-world knowledge to bring to the solution, either for formulation or for interpreting the result sensibly. This is of course possible – there are real-world contexts that any one of us knows nothing about – but the contexts were chosen by an experienced teacher to be within the direct and indirect experiences of Year 10 students. A second possibility is that students have some knowledge of the context, but they focus on the wrong part of that knowledge for the formulation or interpretion of results. For example, some student interviews showed that a familiar context was examined only from a personal point of view, ignoring the mathematical point of view. When judging whether an increase in train ticket prices should be reported on TV as very large, these students considered only the absolute increase (thinking whether it was "large" or not), and ignored the fact that the relative increase was a very small percentage in line with normal inflation, even though they would probably be able to calculate it. A third possibility is that some students did not consider their own knowledge of the context as a source of information, lacking confidence to use it or perhaps believing that common sense should not be used in a mathematics test. One interviewee commented (on a parallel item about increase in sleeping hours): "I didn't quite get if you were supposed to actually use any maths equations to solve it, or if it was just pure logic thinking or anything" (Almuna, 2020, p. C-3).

These research findings may be particularly helpful with respect to how real-world problem context may be treated in the mathematics classroom by teachers and students. The results presented in Table 4.1 indicate that still greater efforts are needed in classrooms, and specifically that teachers in Year 10 classes broaden their perspective to look beyond zero-level and first-level use of context problems to assist students to develop the ability to use the context of a problem thoughtfully and hence use their mathematical knowledge in their lives beyond school. However, to achieve this outcome, the teaching issue of carefully constructing problems that require real-world considerations needs to be addressed. This issue is explored next.

6 Using Level of Context Use as an Analytical Tool to Construct Mathematical Problems

A careful process devised for the Almuna (2020) study helped to guarantee the quality of the process of changing contexts to study the effect of context features on students' performance. The task was to change the context of a problem, whilst controlling – to the best extent possible – for all its other properties. Although it seems obvious, the initial step was to introduce a clearer understanding of the key constructs that played a role in the construction of problems: matching mathematical core, stimulus (i.e., format, number of words, readability, grammar etc), and PISA mathematical competencies. Additionally, the process required making the problems suitable for the Year 10 students in today's society and reflecting purposeful use of mathematics in real-world contexts fitting the PISA construct of mathematical literacy.

This process revealed deep aspects about how context is embedded in the problem-solving process, supporting the observation of Sawatzki (2014) that finding an appropriate new context depends upon imagination, understanding of students' interests, knowledge of students' context familiarity and context engagement, and students' cultural background. Changing the context of a problem introduced the need for students to make new assumptions and draw on different knowledge, especially for second-order use of context problems. For example, consider changing the context of the modified Rock Concert PISA Problem (Figure 4.2) to a classical concert. For a classical concert, it is easier for students to have a mental picture, from personal or vicarious experiences, of a stationary seated audience almost organised in the rows and columns of an array. This perhaps brings a useful multiplicative model to the fore, more than happens for the rock concert context where we might visualise the audience dancing and crowding the stage. With a change of context, students have to model different entities: in this case, seat and aisle rather than body size and spacing. As a result, very careful considerations are needed to construct the parallel items at the second-order level of context use. It was an extremely difficult, although interesting, process.

As another example, Robberies (OECD, 2006, p. 34) asks students to comment on an extract of a newspaper report with a potentially misleading bar graph with a truncated vertical axis. The visual image of the truncated bars supports the newspaper's claim that robberies have hugely increased, but reading the scale shows only a small increase from 508 to 516 robberies in a year. The mathematical core related to considering both absolute and relative change. However, whether a change is truly large or small and whether absolute or relative changes are most important varies markedly with context (Almuna & Stacey, 2014); consider, for example, increased train fares, house prices, change in atmospheric carbon dioxide or the average temperature of the earth.

For the other levels of context use, a change of context could make a problem look different from the original version, but the assumptions and information required in both versions did not alter significantly between the problems.

For example, consider the two problems in Figure 4.6 which were created as variants of Seal problem (OECD, 2009b, p. 117).

These problems are set in different real-world contexts (a "fly in fly out" work roster common in Australia and an airport shuttle bus service). The problems can be solved in mutiple ways, but always somehow involve the mathematical concept of periodicity. The problems are carefully matched. Information is given about the cycles (e.g., roster cycle of 15 + 20 days), but not all the times involved (e.g., time to fly home). Students needed to grasp that a full cycle takes 35 units (i.e., days or minutes) and then begins again. The problems involve the same mathematical content, the same main mathematical competency – reasoning and argumentation (OECD, 2015) – and have closely matched stimuli and answer choices. Both problems require students to translate information into mathematical terms, devise a model to represent the periodic nature of the situation, and extend the pattern to match the result with one of the given options. Both problems are classified as first-level of context use. Each context provides an opportunity for students to interpret and translate information mathematically. For example, for Mine Schedule, students needed to appreciate that (i) the remote location requires Martin to live temporarily at the mine site, (ii) the cycle allows enough time for Martin to be flown back home for his rest period, and (iii) the roster is regular.

Creating matching problems was challenging, even for these first-level of context use problems. Changed contexts changed the units and wording (finally achieving 123 and 121 words respectively). More importantly, Bus Schedule requires a conversion from hours to minutes, whereas Mine Schedule requires conversion from months (and half a month) to days. In order to ensure that the problems had the same mathematical core and drew on similar real-world knowledge, the Mine Schedule problem had to include the (slightly) unrealistic instruction to consider that each month has 30 days. Even in these two situations which are so mathematically close, different contexts influence the problem solving in different ways.

7 Conclusion

This chapter has been motivated by the desire to investigate mathematical literacy, and especially the way in which students deal with real-world contexts in mathematics assessments. A major contribution of PISA has been to draw attention to the importance of students being able to use their mathematics in all aspects of life beyond school, and incidentally, PISA has also drawn attention to the need for mathematics educators to understand better the inter-connection of mathematics and context.

The chapter reported on conceptual work to better understand context, and explained a new scheme to classify problems according to their level of context use. The empirical work demonstrated that the level of context use required to solve a problem is a large contributor to problem difficulty, and that second-order level of context use problems are especially difficult. However, these problems are characteristic of the way in which mathematics is used in problems in real life. The OECD (2009a) notes that "solving this type of problem is the essence of real-world problem solving" (p. 31). Interview components of the reported work point to the need for more attention to these problems in schools, especially as it is likely that students often have a lot of real-world knowledge that they can contribute to solving a problem, but they do not feel confident or permitted to do so. The authors hypothesise that this is the case not only for short PISA-like problems, but also for the more substantial investigations that students would encounter when learning mathematical modelling.

In order to assist teachers in adding a variety of real-world problems to their teaching, the construction of some problems was discussed. The examples provided led to the observation that context is a feature that strongly impacts problem construction both for PISA problems and short or long classroom problems, whenever its author intends to be true to that context. An incorporation or change of context for a mathematical problem produces, for example, unavoidable variation in the number of words, syntax, vocabulary and layout. At the same time, a context can frequently impact on the mathematical core of the problem(s) as discussed in the above section. Added to this, problem developers need imagination, understanding of students' interests, knowledge of context familiarity and context engagement, and students' cultural background. Therefore, it is worth bearing in mind that problem designers, assessors, and teachers face complex issues when writing problems embedded in real-world contexts. The contribution of this chapter has been to unpack these complex issues by providing a deeper understanding of level of context use, and the factors that contribute to it.

Notes

1 The PISA 2022 framework was not available at the time of writing, hence the authors used the framework definitions used for the PISA surveys 2012, 2015, and 2018.

2 There are minor changes to the definition of mathematical literacy and changes to the graphics for the PISA 2022 survey. For more details see the PISA 2022 framework available on https://doi.org/10.1787/dfe0bf9c-en
3 This is an adaptation of an original work by the OECD. The arguments employed in this adaptation are the sole responsibility of the main author of this chapter and should not be reported as representing the official views of the OECD or of its member countries.
4 Reproduced with the permission from the OECD.
5 Reproduced with the permission from the OECD.

References

Almuna, F. (2010). *Investigating the effect of item-context on students' performance on mathematics items* [Master's thesis, The University of Melbourne]. University of Melbourne's Institutional Repository. http://hdl.handle.net/11343/36236

Almuna, F. (2016). Investigating the impact of context on students' performance. In B. White, M. Chinnappan, & S. Trenholm (Eds.), *Proceedings of the 39th annual conference of the Mathematics Education Research Group of Australasia* (pp. 102–109). MERGA. https://files.eric.ed.gov/fulltext/ED572407.pdf

Almuna, F. (2020). *Investigating the effect of mathematics problem context on the performance of Year 10 students* [Doctoral dissertation, The University of Melbourne]. University of Melbourne's Institutional Repository. https://minerva-access.unimelb.edu.au/handle/11343/239230

Almuna, F., & Stacey, K. (2014). Item context factors affecting students' performance on mathematics items. In J. Anderson, M. Cavanagh, & A. Prescott (Eds.), *Proceedings of the 37th annual conference of the Mathematics Education Research Group of Australasia* (pp. 55–62). MERGA. https://files.eric.ed.gov/fulltext/ED572560.pdf

Australian Curriculum, Assessment and Reporting Authority. (2016). *Australian curriculum: General mathematics.* https://www.australiancurriculum.edu.au/umbraco/Surface/Download/Pdf?subject=General%20Mathematics&type=SS

Blum, W. (2002). ICMI Study 14: Applications and modelling in mathematics education. *Educational Studies in Mathematics, 51*(1/2), 149–171. https://doi.org/10.1023/A:1022435827400

Brown, J. P. (2019). Real-world task context: Meanings and roles. In G. Stillman & J. Brown (Eds.), *Lines of inquiry in mathematical modelling research in education.* ICME-13 Monographs. Springer. https://doi.org/10.1007/978-3-030-14931-4_4

Crisp, V., & Greatorex, J. (2022). The appliance of science: Exploring the use of context in reformed GCSE science examinations. *Assessment in Education: Principles, Policy & Practice,* 29(6), 689–710, https://doi.org/10.1080/0969594X.2022.2156980

de Lange, J. (1979). Contextuele problemen [Contextual problems]. *Euclides, 55,* 50–60.

de Lange, J. (1999). *Framework for classroom assessment in mathematics*. NICLA/WCER. http://www.fi.uu.nl/publicaties/literatuur/6279.pdf

Finnish National Agency for Education. (2021). *Subjects at general upper secondary: Maths*. https://www.oph.fi/en/education-and-qualifications/subjects-general-upper-secondary

Greatorex, J. (2014). Context in mathematics questions. *Research Matters, 17*, 18–23. http://www.cambridgeassessment.org.uk/Images/164660-research-matters-17-january-2014.pdf

McCullagh, P., & Nelder, J. A. (1989). *Generalized linear models* (2nd ed.). Chapman and Hall. https://doi.org/10.1201/9780203753736

Ministry of Education of Chile. (2016). *Matemática: Programa de estudio primero medio*. https://www.curriculumnacional.cl/614/articles-34359_programa.pdf

Niss, M., Blum, W., & Galbraith, P. (2007). Introduction. In W. Blum, P. Galbraith, H. Henn, & M. Niss (Eds.), *Modelling and applications in mathematics education: The 14th ICMI study* (pp. 3–32). Springer. https://doi.org/10.1007/978-0-387-29822-1_1

Organisation for Economic Co-Operation and Development. (2006). *PISA released items-mathematics*. http://www.oecd.org/pisa/38709418.pdf

Organisation for Economic Co-Operation and Development. (2009a). *Learning mathematics for life: A perspective from PISA*. OECD Publishing.

Organisation for Economic Co-Operation and Development. (2009b). *PISA 2009 assessment framework. Key competencies in reading, mathematics and science*. OECD Publishing.

Organisation for Economic Co-Operation and Development. (2013). *PISA 2012 results: What students know and can do* (Vol. I). OECD Publishing.

Organisation for Economic Co-Operation and Development. (2015). *PISA 2015 key findings for Australia*. https://www.oecd.org/australia/pisa-2015-australia.html

Organisation for Economic Co-Operation and Development. (2019). *PISA 2018 Assessment and analytical framework*. OECD Publishing.

Organisation for Economic Co-Operation and Development. (2023). *PISA 2022 Assessment and analytical framework*. OECD Publishing. https://doi.org/10.1787/dfe0bf9c-en

Ross, S. M., McCormick, D., & Krisak, N. (1986). Adapting the thematic context of mathematical problems to student interests: Individualized versus group-based strategies. *Journal of Educational Research, 79*(4), 245–252. https://www.jstor.org/stable/27540205

Sawatzki, C. (2014). *Connecting social and mathematical thinking: Using financial dilemmas to explore children's financial problem-solving and decision-making* [Doctoral dissertation, Monash University]. Monash University's Research Repository. (https://doi.org/10.4225/03/58ae5e075ec5e)

Sawatzki, C. (2017). Lessons in financial literacy task design: Authentic, imaginable, useful. *Mathematics Education Research Journal, 29*(1), 25–43. https://doi.org/10.1007/s13394-016-0184-0

Silver, E. A. (1981). Recall of mathematical problem information: Solving related problems. *Journal for Research in Mathematics Education, 12*(1), 54–64. https://doi.org/10.2307/748658

Stacey, K. (2015). The real world and the mathematical world. In K. Stacey & R. Turner (Eds.), *Assessing mathematical literacy. The PISA experience* (pp. 57–84). Springer International Publishing. https://doi.org/10.1007/978-3-319-10121-7_3

Stacey, K., & Turner, R. (2015). The evolution and key concepts of the PISA Mathematics frameworks. In K. Stacey & R. Turner (Eds.), *Assessing mathematical literacy. The PISA experience* (pp. 5–33). Springer International Publishing. https://doi.org/10.1007/978-3-319-10121-7_1

Stillman, G. (1998). The emperor's new clothes? Teaching and assessment of mathematical applications at the senior level. In P. Galbraith, W. Blum, G. Booker, & D. Huntley (Eds.), *Mathematical modelling: Teaching and assessment in a technology-rich world* (pp. 243–253). Horwood Publishing.

Treffers, A. (1987). *Three dimensions: A model of goal and theory description in mathematics instruction.* Reidel. https://doi.org/10.1007/978-94-009-3707-9

Turner, R., & Adams, R. (2012, April 13–17). *Some drivers of test item difficulty in mathematics: An analysis of the competency rubric* [Paper presentation]. Annual Meeting of the American Educational Research Association (AERA), Vancouver. https://research.acer.edu.au/pisa/7/

van den Heuvel-Panhuizen, M. (2005). The role of contexts in assessment problems in mathematics. *For the Learning of Mathematics, 25*(2), 2–23. https://flm-journal.org/Articles/1957A5517A35A04A1E9F6310B923E0.pdf

Watson, R., & Yip, P. (2011). How many were there when it mattered? *Significance, 8*(3), 104–107. https://doi.org/10.1111/j.1740-9713.2011.00502.x

CHAPTER 5

Interplay between Contextual and Mathematical Literacy in Task Design

The Math-LIGHT Program as an Example

Roza Leikin, Masha Boriskovsky and Math-LIGHT team

Abstract

In this chapter we describe and exemplify principles of the design of tasks in the Math-LIGHT (Mathematical Literacy Insights for Growing Horizons Tasks) program. The Math-LIGHT program is directed at the development and support of teachers' proficiency in navigating mathematical instruction aimed at developing students' mathematical literacy and broadening their horizons in fields such as natural and social sciences, arts, music and sports. We draw a distinction between Mathematical Literacy and Contextual Literacy, assuming that both are essential for successful completion of modelling tasks. We describe the Stepped Structure of mathematical tasks employed in Math-LIGHT, a structure which allows their implementation with students of different levels of mathematical and contextual literacy. Additionally, the chapter presents expert teachers' conceptions about the role of integration of mathematical and contextual literacy in the implementation of Math-LIGHT in middle school classrooms.

Keywords

mathematical literacy – task design – stepped tasks – Math-LIGHT program

1 Introduction

We start this chapter with a short problem prepared for publication in an electronic journal devoted to the popularization of literacy problems in school mathematics. We invite readers to solve the problem, presented in Figure 5.1, before reading the rest of this chapter.

Since this chapter is included in a volume devoted to learning for mathematical literacy, we assume that most readers will solve this problem immediately. Our experience with this problem (asking dozens of colleagues and

 | DOI:10.1163/9789004710962_005

Eran and Michal look at the thermometer when going outside.

Michal goes outside shopping while the temperature outside is 25 degrees. She wears a silk dress without sleeves and a straw hat with a flower.

Eran goes outside with his dog while temperature outside his home is 25 degrees. He wears a warm coat and wool cap with a pompon.

How can this happen? Explain your answer.

FIGURE 5.1 Hat problem

friends) is that only about one out of five people answer this question immediately. Those who do not answer the question themselves are usually excited about the answer; however, they claim that this problem is not mathematical. Using the definition of literacy as a person's knowledge in a particular subject or field,[1] we ask: "What kinds of literacy are essential to solve the Hat Problem?", and return to the discussion of the related literacy components in the concluding section of the chapter.

In analogy with the distinction drawn between content and context in mathematical problem solving (Caldwell, 1979; Goldin & McClintock, 1979; Kulm, 1979; Webb, 1979), we draw a distinction between mathematical literacy and contextual literacy. Mathematical literacy is associated with the mathematical content to be used when solving the problem, while contextual literacy is linked to the phenomena or situation described in the problem. We use the terms "conceptual literacy" and "mathematical literacy" interchangeably by including solvers' knowledge and skills linked to mathematical concepts, properties, algorithms, and problem-solving strategies used when solving the problem. Contextual literacy includes personal acquaintance with different fields (cf. the above definition) such as, for example, arts, natural or social sciences, engineering and technology. The knowledge of mathematics related to a particular context is only a part of contextual literacy, which can include the history of the field's development, its uses or its contribution to humankind. On the other hand, mathematical literacy can include knowledge of different contexts in which a particular concept or property can be used. The term "contextual literacy", in contrast to "general literacy", acknowledges differences between people's acquaintance with specific context areas. Both types of literacy (mathematical and contextual) are related to an individual's ability to obtain, understand, and use information in accordance with specific goals. Both kinds of literacy can be developed in individuals through educational activities.

In this chapter, we discuss the Math-LIGHT program, a program that integrates contextual and conceptual literacy aimed at the development of literacy of both kinds: the tasks are aimed at developing students' mathematical competencies by acquainting them with various phenomena in science and life and broadening their contextual literacy through the solving of mathematical problems. One of the major principles of Math-LIGHT design is allowing all students to engage in solving the tasks independently, regardless of their level of mathematical and contextual knowledge. This goal is attained by using a Stepped-Tasks design (Leikin, 2019, Leikin & Ovodenko, 2021) which comprises sets of problems organised in a top-down structure. The three tasks presented in this chapter illustrate possible ways of stepping problems by decreasing their conceptual and contextual complexity. Along with the distinction between mathematical and contextual literacy, Stepped-Tasks' development integrates variations in conceptual and contextual density.

In what follows, we analyse the design and implementation of the Math-LIGHT tasks. We provide a short literature review on the topic of literacy and present examples of three Math-LIGHT units that explain and illustrate the stepped structure of the intended mathematical activities. We also briefly discuss the implementation of the Math-LIGHT units in schools and analyse teachers' conceptions about the interplay between contextual and mathematical literacy.

2 Background

2.1 *Mathematical Literacy and Contextual Literacy*

There are multiple definitions of literacy on the internet and in the educational literature. The definition presented above applies to literacy in different fields. In contrast, the UK National Literacy Trust (2017) defines literacy as "an ability to read, write, speak and listen in a way that lets us communicate effectively and make sense of the world" ("What is literacy" section). In association to literacy, reading and writing are considered tools for either understanding or acquisition of different contexts (McKenna & Robinson, 1990). The complexity of literacy is also reflected in the multiple terms used in the literature: computer literacy, cultural literacy, media literacy, digital literacy and internet literacy or distinction between literacy and numeracy (Goos et al., 2013). Literacy of everyday life – at home, in the workplace and at social events – is significantly different from school-type literacy (Mikulecky in Venezky et al., 1990). An additional variation to the concept of literacy can be seen in the discrepancy between "functional literacy", which is connected to attaining concrete goals, and the levels of literacy that any individual needs to attain for a specific goal (Collins & O'Brien, 2011; Wallendorf, 2001). The development of the concept of literacy over time can be seen in OECD documents related to the Organisation for Economic Co-operation and Development's (OECD) Programme for International Student Assessment (PISA assessment) (OECD, 2010, 2022).

The PISA 2022 Mathematics Framework defines mathematical literacy in close connection to mathematical proficiency. Mathematical literacy is:

> an individual's capacity to reason mathematically and to formulate, employ, and interpret mathematics to solve problems in a variety of real-world contexts. It includes concepts, procedures, facts and tools to describe, explain and predict phenomena. It assists individuals to know the role that mathematics plays in the world and to make the well-founded judgments and decisions needed by constructive, engaged and reflective 21st century citizens. (OECD, 2022, p. 8)

Attention is paid to literacy in different fields, with an emphasis on reading and writing literacy, mathematical literacy, financial and statistical literacy. According to the OECD (2022), mathematics literacy requires understanding of content in four major branches of mathematics: space and shape, quantity, uncertainty and data, and change and relationship. The importance of mathematical literacy in the 21st century is related to the huge amount of quantitative data encountered in different contexts and represented in different ways that require information processing, logical inferences, critical evaluation of the situation and interpretation (De Lange, 1996; Schoenfeld, 1992). Researchers in the UK, Canada, South Africa, Australia, and New Zealand use the term "numeracy" to denote skills related to the effective use of mathematics in different contexts (Geiger et al., 2015; Goos et al., 2013, 2023). Numeracy has a similar meaning to quantitative literacy (Steen, 2001) or mathematical literacy (OECD, 2010).

Mathematical literacy per se has a variety of different definitions. According to Julie (2006), definitions of mathematical literacy constitute a continuum of meaning: "At the one end of the continuum, Mathematical Literacy is viewed as entry into Mathematics and at the other end as a means to interact with mathematical installations in society" (p. 62). Kilpatrick (2001) stressed that the terms "mathematical literacy" and "mathematical proficiency" are very close to each other as both include conceptual understanding, procedural fluency, strategic competence, adaptive reasoning, and productive dispositions.

Furthermore, continuously developing literacy in a specific field is an indicator of proficiency in this field that is rooted in goal directed practice (Street, 2003, 2006; Wallendorf, 2001).

The distinction between contextual and mathematical literacy in the mathematical modelling cycle (Figure 5.2), in our view, is analogous to the distinction between content and context in mathematical problem solving (cf., Goldin & McClintock, 1979; Kulm, 1979; Webb, 1979). As noted above, we prefer using the term contextual literacy instead of "real world" (e.g., Kaiser & Stender, 2013) or "extramathematics" (e.g., Borromeo Ferri, 2006), since in our view, as explained above, contextual literacy includes mathematical elements while mathematical literacy can be purely mathematical or connected to a particular context (e.g., according to the OECD definition). The importance of this distinction is also related to the fact that identical mathematical models and results can be related to different mathematical contexts, and without knowledge of the specific context one cannot create a mathematical model (we elaborate these ideas in the description and analysis of Math-LIGHT Tasks). Note that while, according to Kaiser and Sriraman (2006), modelling integrates pragmatic (cf., Pollak, 1968) and scientific-humanistic (cf. Freudenthal, 1968) goals. This integration of goals is linked in our view to the integration of mathematical and contextual literacy in the modelling tasks.

As noted above, the complexity of mathematical modelling is related to integration in different contexts such as personal, occupational, societal, scientific, cultural, historical, and economical. Thus, modelling activities can require students to be literate in a variety of contexts. Integration of mathematical and contextual literacy can contribute to the advancement of mathematical

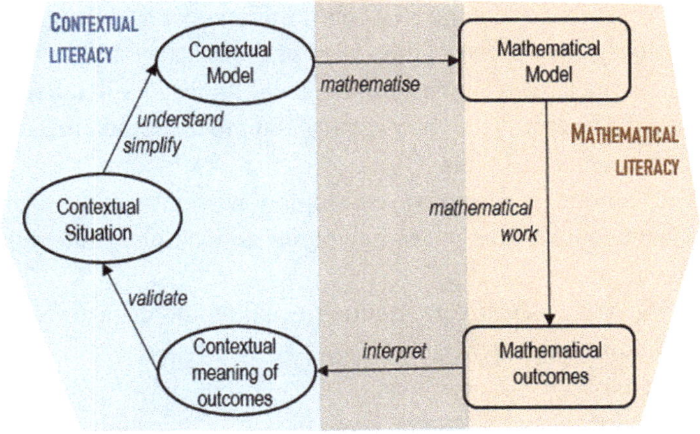

FIGURE 5.2 Modelling cycle from educational modelling (based on Kaiser & Stender, 2013; the term "real life" is replaced by "context" by the author)

thinking and broaden the horizons of both teachers and students. The main obstacles that teachers face when implementing modelling tasks are a lack of time, a lack of teaching materials and difficulties with the evaluation of these tasks (Borromeo Ferri & Blum, 2014). There is a need for additional studies on instructional practices and teachers' proficiency focusing on teaching with modelling tasks and related classroom activities, as well as on the connections between the above and student achievements (Kaiser et al., 2017).

We argue that mathematical literacy tasks do not necessarily include components of the contextual literacy zone of the modelling cycle (see Figure 5.2). Nevertheless, mathematical literacy clearly connects mathematical content with specific context, and thus, solving literacy tasks develops both mathematical and contextual literacy. To illustrate these ideas, we describe and exemplify tasks from the Math-LIGHT program and analyse teachers' conceptions related to the implementation of the program in middle school.

3 Math-LIGHT Program and Units

The Math-LIGHT (Mathematical Literacy Insights for Growing Horizons Tasks) program is directed at the development of mathematical literacy and broadening students' horizons in different contexts. The tasks are usually situated in the Mathematical Literacy Zone (Figure 5.2) while the instructional designers work in the Contextual Literacy Zone (Figure 5.2) to design meaningful and approachable-for-students tasks. The program has a set of cognitive and socio-emotional goals.

- Cognitive goals:
  - Context-related goals: developing students' awareness and understanding of the phenomena or process at the basis of the task.
  - Content-related goals: developing mathematical knowledge related to the task context; developing problem-solving and proof competencies as well as other mathematical skills.
  - Integration of content and context: developing awareness and understanding of connections between mathematics and various contextual phenomena and processes.
  - Domain-general goals: developing mental flexibility and critical thinking, self-regulation and creativity.
- Socio-emotional goals:
  - Motivational goals: raising students' awareness of the usefulness and necessity of mathematics in different contexts and fields.

- Orientation goals: developing genuine interest in mathematics and getting a glimpse of mathematically-rich professions; developing social responsibility.
- Communication goals: advancing collaborative and communicative skills through work on the tasks and discussion of the solution processes and outcomes, encouraging argumentation, explanation and justification.

In terms of the mathematical modelling cycle, the connection between the contextual situation and the corresponding contextual model is made by the Math-LIGHT team[2] of task designers. Students are usually introduced to the contextual model and asked to design a mathematical model and solve the problem. In some cases, we explicitly describe the connection between the contextual and conceptual models and ask students to work with a mathematical model and interpret the outcomes in terms of the context.

The Math-LIGHT program creates an opportunity to cultivate 21st century skills, including mental flexibility and openness to the ideas of others, critical thinking and meta-cognitive thinking, self-regulation and collaboration, creativity and curiosity. The program promotes students in broadening their common knowledge in different fields, including (but not limited to): economics, architecture, science, and society. In order to enhance teachers' proficiency and provide them with tools that will assist in the development of mathematical literacy in students, the Math-LIGHT program provides teachers with specially designed tasks that enable them to advance students' contextual and mathematical literacy. The program developers hope that Math-LIGHT will develop genuine curiosity about the connections between mathematics and different fields of life among both students and teachers.

In the Math-LIGHT units one contextual situation can serve as the basis for a set of problems related to a particular mathematical topic (illustrated in Unit 1 and Unit 3 below). Alternatively, a situation can be the basis for a set of problems involving different mathematical topics, all related to a particular context (illustrated in Unit 2 below). The program encourages the application of mathematical knowledge and skills through the analysis and characterisation of situations in different fields. It is available online and is linked to the school curriculum while also extending content beyond the curriculum's borders. The majority of the Units include dynamic explorative applets designed in GeoGebra which are directed at better understanding of the contextual situation and related mathematical concepts. In order to develop students' deeper understanding of content and context and to encourage self-regulated learning, the units are designed in the form of Stepped Tasks as described below.

3.1 *Stepped Tasks*

Stepped Tasks are sets of problems designed in a top-down structure that enables the problem-solving process to be regulated by students through the variation of mathematical challenge according to students' needs (Leikin, 2019; Leikin & Ovodenko, 2021). A Stepped Task starts with a complex problem, called the "target problem", which is accompanied by paths that include steps of different levels of challenge. Each step includes a number of problems with a reduced level of challenge with respect to the previous step(s). The problems in a Stepped Task belong to a particular mathematical topic related to the contextual situation. Pure mathematical Stepped Tasks are designed through decreasing the level of conceptual density of a problem (i.e., the number of concepts and their properties needed to solve the problem; cf. Silver & Zawodjewsky, 1997). Stepping of Math-LIGHT units is performed by decreasing either conceptual or contextual density. Each step in a Stepped Task includes problems that share a similar level of complexity. We use the term Varying conceptual (mathematical) and contextual challenge to refer to decreasing the level of challenge that a task entails while ensuring that the task remains challenging for the students (Leikin, 2019; Leikin & Ovodenko, 2021). Stepped Tasks' design supports teachers' capacity to conduct complex problem solving in the classroom by varying the level of challenge embedded in the tasks. In addition to the variations in contextual and conceptual density, Stepped Tasks' design is based on changes to the length of the solution or logical chains in a proof and to the level of knowledge and skills required to solve the problem.

The target problem is explicitly presented to all the students, and they are instructed to solve the problem. The students may use the steps according to their own need to decrease the level of challenge. As such, Stepped Tasks can be described as a goal-directed, student-regulated activity. In this way the Math-LIGHT program provides multiple opportunities to match the challenge to students' level of literacy. In practice, we observed different modes of implementation of the tasks in mathematics classes. Here we distinguish between the top-down and bottom-up structures of implementation: in the latter, teachers open the unit with the easiest problems and gradually raise the level of complexity. The tasks' design thus provides teachers with opportunities for flexible lesson management.

4 Examples of Math-LIGHT Tasks and Analysis of Their Complexity

In this section we present three Math-LIGHT tasks. The tasks are analysed according to the knowledge needed to solve the problems: general literacy knowledge, mathematical literacy and mathematics knowledge and skills.

Unit 1: Different Temperature Scales

In Israel, the standard way of measuring temperature is in Celsius scale degrees, °C.
In the United States, the standard way of measuring temperature is in Fahrenheit scale degrees, °F.
In science, the standard way of measuring temperature is in Kelvin scale, K.
Picture 1 depicts three thermometers that measure the temperature in Kelvin (K), Celsius (°C) and Fahrenheit (°F) scales.

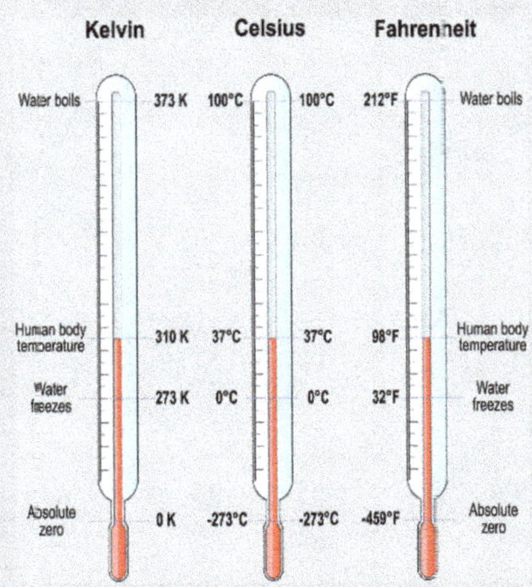

PICTURE 1[3]

Target Problem 1

Absolute zero is equal to 0K. Absolute zero is the lowest temperature that exists. Material cannot be cooled to absolute zero, but it is possible to approach absolute zero. The lowest temperature possible was defined by the British researcher William Thompson, known as Lord Kelvin, in 1848.

Look at the Picture 1 that depicts thermometers and answer the following questions. For each question, explain your solution in detail. Find an additional explanation/solution.

a. What is absolute zero in degrees Celsius? In degrees Fahrenheit? Find functions representing the relationships between the Kelvin and Fahrenheit temperature scales.
b. Does the temperature −400°F exist?
c. Does the temperature −500°F exist?

Solution

a. −273°C = $0K$, −273°C = −459.4°F, 0K = −459.4°F
b. The function representing the relationship between the Kelvin and Fahrenheit scales is: $y_F = 1.8x_K - 459$.

Step 1.1

Problem 1.1.1 For each one of the graphs of functions A, B, C and D, determine which relationships between pairs of temperature scales (*C*, *F*, and *K*) are represented. Explain your answers in as many ways as possible. If needed, use the dynamic applet to explore the relationships.

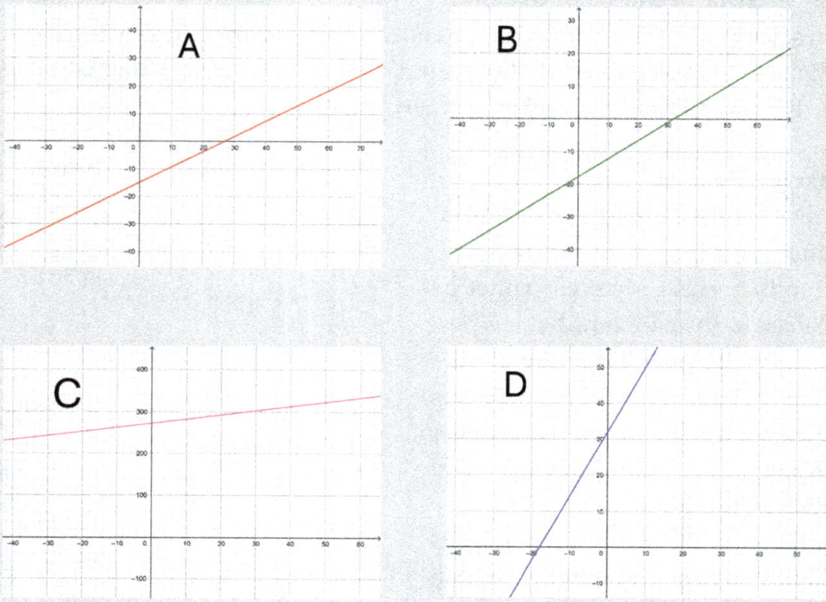

Solution 1.1.1

Graph A: N/A

Graph B: $F \rightarrow C$

Graph C: $C \rightarrow K$

Graph D: $C \rightarrow F$

Step 1.2

Look at the pictures of the thermometers (Picture 1) and solve problems 1.2.1, 1.2.2

Problem 1.2.1

a. How many degrees Fahrenheit is the temperature 0°C?
b. How many degrees Fahrenheit is the temperature 140°C?

Problem 1.2.2

a. If the temperature increases by 10K, what was the increase in degrees Celsius?
b. If the temperature increases by 10K, what was the increase in degrees Fahrenheit?

Explain your solutions in as many ways as possible.

Solution 1.2.1
A temperature of 140°C is equal to 284°F.
Solution 1.2.2 An increase in temperature of 10K equals an increase of 10°C, an increase in temperature of 10°C is equal to an increase of 18°F, and so an increase in temperature of 10K is equal to an increase of 18°F. (Picture 1).

Step 1.3
Look at the pictures of the thermometers (Pictures 1) and solve Problem 1.3.1

Problem 1.3.1
Complete the table below.

Temperature in °C	Temperature in K	Temperature in °F
0°C		
10°C		
20°C		

You are advised to use the corresponding applet.

Solution 1.3.1

Temperature in °C	Temperature in K	Temperature in °F
0°C	273 K	32°F
10°C	283 K	50°F
20°C	293 K	68°F

TABLE 5.1 Analysis of target problem 1

Contextual (physics) literacy	– Measuring temperature in 3 scales/units: Celsius, Kelvin and Fahrenheit – Reading temperature from a thermometer/ Comparing different scales
Mathematical literacy	– Using a linear function to explore and understand relationships between pairs of different scales.

Mathematics knowledge and skills	– Constructing linear relationship / linear function
	  – Finding coefficient of linear relationship, finding equations of the functions
	  – Translation between different representations: pictorial – graphical – symbolic
	  – Drawing graphs of the functions

Target Problem 1 is a complex problem with a high level of contextual and conceptual density. It simultaneously requires the literacy, knowledge, and skills presented in Table 5.1.

If the task is too complex for students, we provide them with steps constructed in a top-down structure. The stepped structure of the task allows students to solve problems with reduced conceptual density, either through self-regulated decision-making about use of the steps or via a teacher's recommendation. The conceptual density of Problem 1 is reduced in Step 1.1 since the requirement to construct graphs of functions is removed. In Step 1.2 the conceptual density of Problem 1.1.1 is reduced since the problems are focused on specific points on the graphs or on the scales in Pictures 1, 2, and 3. Problem 1.3.1 in Step 1.3 draws participants' attention to the thermometer scales and the relationships between the scales, focusing on three temperatures.

Many students and teachers report that the lesson introduced them to different temperature scales; they now see the thermometer differently, and they very much like seeing linear relationships in a real-world situation.

Unit 2: Basketball

The game of basketball is played by two teams, each with five players. The goal of the game is to get more points than the other team. Points are earned by throwing the ball through the basket (hoop) which has a diameter of 45 cm. Throwing the ball into the basket during a game earns a different number of points depending on the position of the player on the court (see Picture 2.1) as follows

3 points if a player throws from or from behind the "three-point line" (Picture 2.2).
2 points inside the "three-point circle"
1 point for a successful penalty

The group with more points at the end of the game wins.

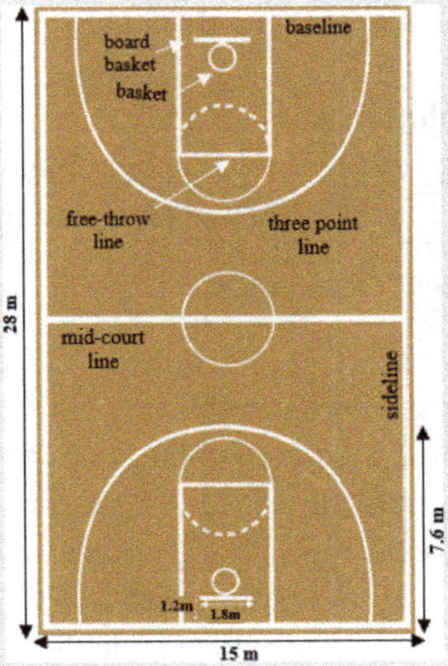

PICTURE 2.1[4]
(the numbers in the figure are approximate)

Target Problem 2.1
Danny plans to throw a 3-point shot.
He's not sure whether to aim for the basket, which has a diameter of 45 cm, from point B or point D. Point B is on the sidelines. Point C is the middle of the mid-court line.
Point D is on the line connecting point C and point O which is under (the projection of) the center of the basket on the court.
The length of DC is 180 cm (see Picture 2.2). Danny prefers to throw from the point that is closest to the center of the basket. Help him to decide which point is closer to the basket.

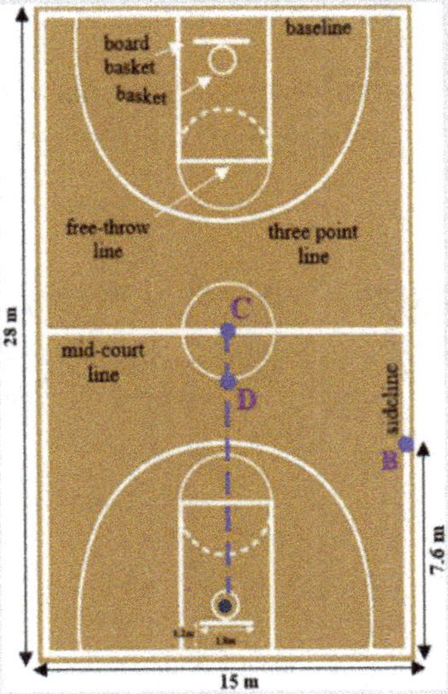

PICTURE 2.2[5]
(the numbers in the figure are approximate)

Explain why comparison of the distances from the points D and B to the center of the basket can be replaced by comparison of the distances from the points D and B on the court to point O.

Solution 2.1

The distance DO is approximately: 14 – 1.8 – 1.2 – 0.2 = 10.8 *m*

Using the Pythagorean Theorem, the distance BO is approximately:

$$\sqrt{(7.6-1.2-0.2)^2+(15:2)^2}=9.7\,m$$

Target Problem 2.2

Shira, a basketball fan, talks to her friends after the game and tells them about Shelly's field goal percentage:

- 70% of Shelly's penalty shots went into the basket, and each of them earned her one point
- 25% of her shots from the three-point line went into the basket and each of them earned her 3 points
- 50% of her other shots went into the basket and each of them earned her 2 points. (Picture 2.3)

How many shots of each type did Shelly score?

Is it possible for two players to earn 18 points, with the same field goal percentage but having a different total number of goals? Explain.

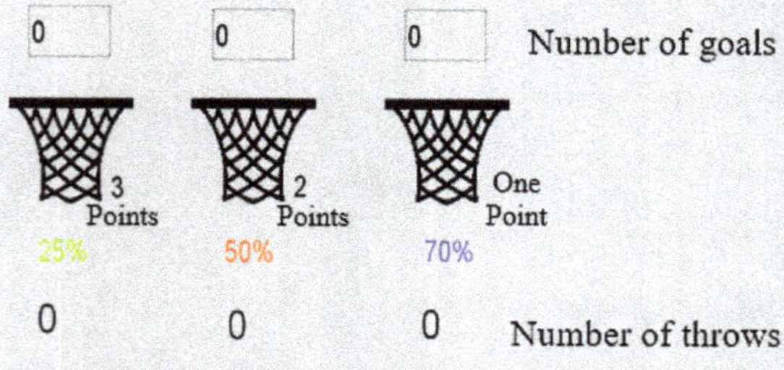

PICTURE 2.3

Solution 2.2

The number of free throws must be divided by 10, each shot is worth a point: Shelly threw 10 shots and scored on 7 of them (7 points in total).

There are 11 points left, which can be a composition of 2- and 3-point throws, and there are two options:

- 4 throws of 2 points each + 1 throw of 3 points (respectively: 8 and 4 throws) or
- 1 throw of 2 points + 3 throws of 3 points each (respectively: 2 and 12 throws)

There are two possibilities:

Possibility 1:	
7 throws of successful penalty	7 points
1 throw of 3 points	3 points
4 throws of 2 points	8 points

Possibility 2:	
7 throws of successful penalty	7 points
3 throws of 3 points	9 points
1 throw of 2 points	2 points

Target Problem 2.3
Shelly throws a ball into the basket and gets 3 points.
The trajectory of the ball is a parabola.
Find the equation of the parabola if Shelly throws the ball from a height of 150 cm when she is on the 3-point circle.

PICTURE 2.3[6]

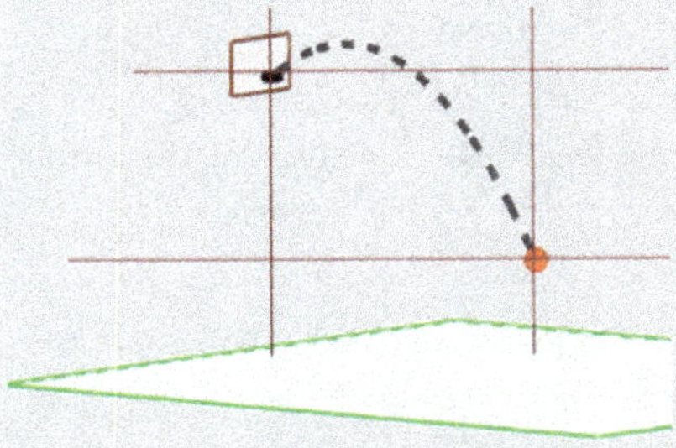

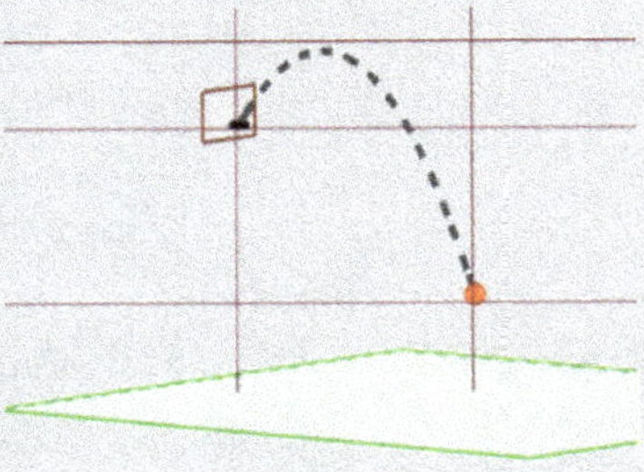

Solution 2.3

The distance between Shelly's legs and the projection of the center of the basket on the floor is 6.2 m (7.6 – 1.2 – 0.2) (*see Unit* 1).
We add a system of axes, see picture 2.4.

The standard representation of a quadratic function is:
$f(x) = ax^2 + bx + c$. The function has a maximum, so $a < 0$.
T is a point on the graph of the function.
Therefore the function is
$f(x) = ax^2 + bx + 3.1$. We set the coordinates of S: (6.2, 1.5):
$1.5 = a \cdot 6.2^2 + b \cdot 6.2 + 3.1$b can be expessed as:

$$b = \frac{-1.6 - 38.1a}{6.2} = \frac{62}{247} - 6.2a$$

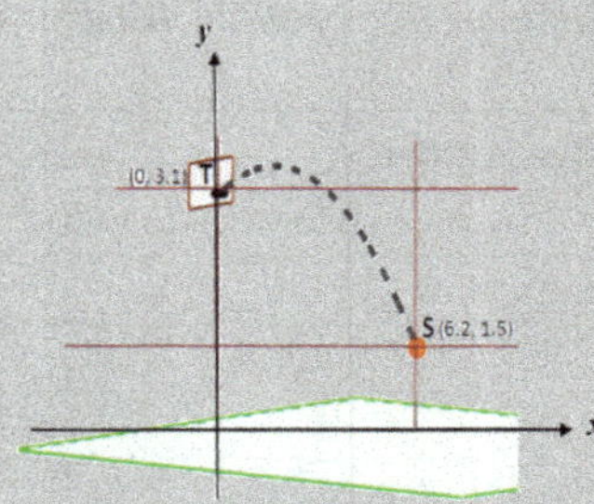

PICTURE 2.4

And we get a family of quadratic functions:

$$f(x) = ax^2 + \left(\frac{62}{247} - 6.2a\right)x + 3.1$$

Unit 2 illustrates how one contextual situation allows students to work with a variety of mathematical content. Basketball is analysed in light of the linear function, Pythagorean Theorem, distances between points on a plane and in space, combinations of numbers depending on players' positions, and the quadratic function. Table 5.2 presents a detailed analysis of the three Target Problems in the unit.

TABLE 5.2 Analysis of target problem 2

Contextual (Sports) literacy	– Basketball rules – Basketball court – Matching answers to the real world (for example: the number of throws is an integer)
Analysis of target problem 2.1	
Mathematical literacy	– Understanding a geometric sketch of a basketball court and translating it into a mathematical model – Switching between different representations of information – Horizontal and vertical lines

(*cont.*)

TABLE 5.2 Analysis of target problem 2 (*cont.*)

Mathematics knowledge and skills	– Right triangle, rectangle – Circle: centre, radius, diameter – Distance between two points on a plane – Distance between the points in space – Pythagorean theorem
Analysis of Target Problem 2.2	
Mathematical Literacy	– Combinatorial reasoning related to successful and unsuccessful throws from different locations on the ground. – Understanding the rules of the game (relationship between throws and goals) and modelling the rules
Mathematics knowledge and skills	– Percentage and its meaning – Linear combinations of numbers $1 \times k + 2 \times n + 3 \times m$
Analysis of Target Problem 2.3	
Mathematical Literacy	– Trajectory of a thrown ball is a parabola – Connecting between the actual trajectory and its sketch
Mathematics knowledge and skills	– Quadratic function and its representations (standard and vertex representations) – Algebraic and graphic meaning of a point located on a graph – Families of quadratic functions – Equation of a quadratic function through 2 points – Solution of quadratic equations

Unit 3: Perspective in Impressionist Painting

In this unit we will learn about perspective in impressionist paintings. Let's consider two paintings.

The first one is **Fishing Boats on the Beach.**

This painting was created in 1888 by the great Dutch painter Vincent van Gogh (1853–1890). We will focus on the part of the painting with 4 boats with white sails.

Try to describe the relationship between the 4 boats.

PICTURE 3.1[7]

The second is *Pont Neuf* (new bridge), *Paris.*

This painting is by prominent French impressionist painter Auguste Renoir (1841–1919) in 1872. We will focus on the part of the bridge with streetlights.

What regularities can be observed?

PICTURE 3.2[8]

Target Problem 3
In the two pictures above perspective lines are drawn.

Based on previous lessons (in which you learned how to draw equal and equidistant objects) answer the following questions:

a. Suppose that the boats are of equal height, are they equidistant?
b. Are the streetlights on the bridge equidistant?

Explain your answer.

PICTURE 3.1A

PICTURE 3.1B

Solution 3
One of the ways of solving:

Construct midlines of the trapezoids and check that the "mid-segment" passes through the point of intersection of the diagonals (which is on the median of the triangle with the vertex at the vanishing point).

The boats are at different distances. Points O1 and O2 are not on the "boat" lines.

The street lights are equidistant. Points O3 and O4 are on the "streetlights" lines.

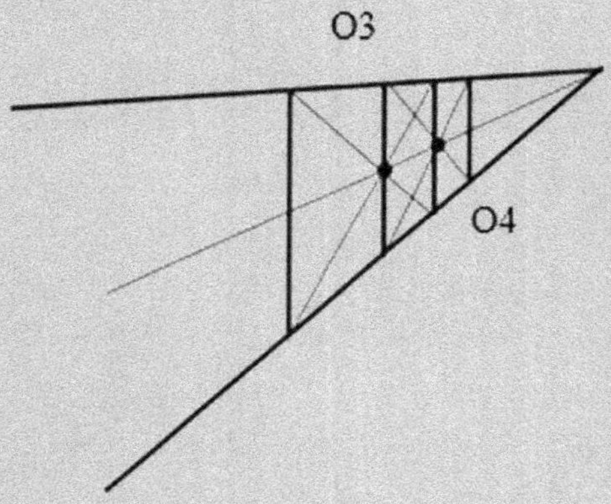

We use the property of equidistant objects in perspective drawings: Of three equal equidistant objects in a perspective drawing the "mid-segment" of the corresponding plane geometry figure is the harmonic mean of the two "side" segments; that is, the mid-segment passes through the point of intersection of the diagonals in the corresponding trapezoid.

A detailed analysis of Target Problem 3 is presented in Table 5.3.

5 Teachers' Conceptions about the Role of Context and Content in Math-LIGHT Units

5.1 *The Setting*

Implementation of the Math-LIGHT program requires teachers to significantly change their teaching practices. They are required to teach problems that

TABLE 5.3 Analysis of target problem 3

Contextual (Arts) literacy	– Perspective, vanishing point – Earth, horizon – Art, paintings, painters, architecture
Mathematical Literacy	– Understanding a geometric sketch related to a perspective painting – Translating painting into a geometry model – Understanding the rules of equal equidistant objects in a painting – Formulation of the contrapositive statement: -from construction to examination[9] – Segment connecting lateral sides of a trapezoid, parallel to bases, through the point of intersection of the diagonals in the trapezoid, represents a real object in the painting – The segment represents an object in the painting (a) equal to the objects represented by the bases of the trapezoid and (b) equidistant from these two objects
Mathematics knowledge and skills	– Parallel and vertical lines (plane and space) – Similar triangles – Midline in trapezoid that connects the bases is on the median of the triangle with its vertex at the vanishing point – Trapezoid – Using construction of equidistant segments to examine equality of distances – Segment connecting lateral sides of a trapezoid, parallel to bases, through point of intersection of the diagonals in the trapezoid – The point of intersection of diagonals bisects the segment – The length is the harmonic mean of the bases

include context that is unfamiliar to them and, moreover, require use of mathematical concepts in an unfamiliar situation. This implies that careful teacher professional development is required. To this end, Math-LIGHT communities of practice were created in which teachers could learn about and practise use of these unique problems. They are asked to report on the implementation of each task by completing a google-form questionnaire. The report form included, among others, three questions which are under consideration in this chapter:

– What role does context play in teaching and learning mathematics with Math-LIGHT tasks?

- Does context contribute to the development of mathematical understanding?
- To what extent does context raise students' interest in learning mathematics?

During the 2020–2021 school year the Math-LIGHT program was implemented in middle school mathematics classes (7th, 8th, and 9th grades). We received a total of 260 reports on implementation of different tasks. Of these reports, 11 presented teachers' views on their experiences with the Temperature Scales Task and 40 reports related to the implementation of the Basketball Task (Units 1 and 2 described earlier in this chapter).

5.2 *Findings*

As presented in Figure 5.3, the majority of teachers felt that the authentic context embedded in the problems helped students' mathematical understanding: 75% of teachers said the context provided in the basketball problem boosted mathematical understanding, while 82% said the same for the temperatures problem. Teachers' perceptions of the role of the authentic context were similar for both problems. In both cases, teachers felt that the main contribution of the context was to provide a framing device for the problem that provoked students' curiosity and interest. The basketball problem was ranked higher here, with 68% of teachers saying that the context aroused students' interest, compared to 55% for the temperature scales problem. The second-ranked role of the authentic context was "connection to everyday life": 48% of teachers felt that the basketball problem helped students connect classroom mathematics to daily life, while 45% said the same about the temperature scales problem. The third role mentioned was drawing connections between mathematics and other fields and clarifying the essentiality of mathematics to other fields: 28% of participating teachers felt that the context in the basketball problem helped to draw these connections, while 27% said the same about the context of the temperature scales problem. Another suggested role of context was broadening students' horizons: regarding both problems, 18% of teachers viewed this role as relevant.

6 Concluding Notes

In this chapter, we introduce a distinction between mathematical literacy and contextual literacy using the mathematical modelling cycle (Figure 5.2). This distinction between mathematical literacy and contextual literacy is made in analogy to the distinction between content and context in mathematical

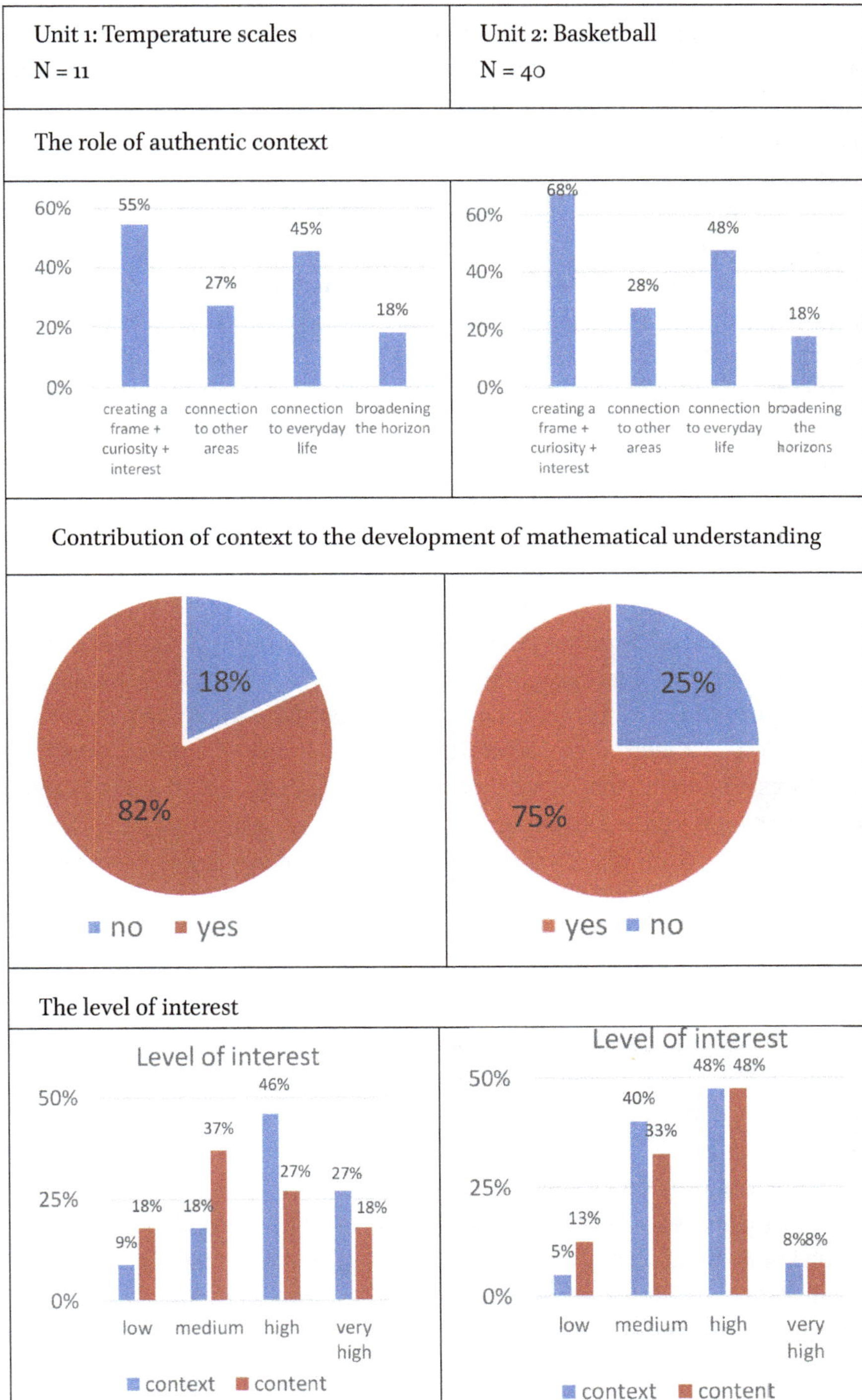

FIGURE 5.3 Findings from teachers' reports

problem solving (Goldin & McClintock, 1979). We argue that contextual literacy is essential for the design of the contextual model and its translation to a mathematical model while mathematical literacy plays a critical role in designing and resolving the mathematical model and interpreting the outcomes in terms of the context.

Math-LIGHT explicitly introduces students to the contextual model and supports them in solving the task by employing a Stepped structure. Moreover, in addition to stepping the tasks by means of reducing the conceptual (i.e., mathematical) density used in pure mathematical tasks (Leikin, 2019), in this chapter we introduce the concept of stepping literacy tasks by decreasing the contextual density by requiring students to use only a partial set of contextual information. For example, in Unit 1 the contextual complexity is decreased by inviting students to consider the relationships between only two different temperature scales instead of considering three scales at once. Math-LIGHT problems engage students with work within the mathematical literacy zone (Figure 5.2) while work within the contextual literacy zone is performed to a large extent by the team of instructional designers.

We describe and illustrate three different types of Math-LIGHT tasks. Unit 1 illustrates stepping conceptual and contextual literacy for similar content and context. Unit 2 illustrates that the same context (basketball) can be used in connection to different mathematical content: linear equations, quadratic functions, and the Pythagorean theorem (in a plane and in 3D space). Unit 3 illustrates how students can use their skills in plane geometry in the context of engagement with art. The decrease of conceptual literacy is performed by using hints about the vanishing points and applying the lines that pass through the top and bottom points of the objects under consideration. The students are asked to apply skills learned in previous lessons.

Our experience with teachers demonstrates that most are highly enthusiastic with respect to the implementation of Math-LIGHT tasks in their classes. Teachers find the additional context both interesting and informative and believe that it inspires students' curiosity and deepens their understanding of the material. However, in many cases the conceptual knowledge is new to the teachers, and they do not feel sufficiently confident to implement the problems in their classes. This is similar to what we found in previous research involving Stepped Tasks. Our research showed that the Stepped Tasks' top-down structure requires adjustments to teachers' practices, and that teachers initially lacked confidence using the tasks. With time, however, teachers gain confidence, and increasingly take advantage of the tasks' top-down structure. We believe that here, too, teachers will gain confidence as they gain experience.

When the thermometers in the two cities show that temperature outside is 25 degrees the following solution is possible:

Michal wears a silk dress and a straw hat with a flower in Tel Aviv.

Eran wears a warm coat and wool cap with a pompom in New York.

FIGURE 5.4 Solution to the Hat problem

Overall, these tasks show high potential for strengthening students' modelling skills and deepening their mathematical and contextual literacy.

Let us return to the Hat problem (Figure 5.1) presented in the introduction to this chapter. When the problem is presented to students it seems to some of them unclear and unapproachable or related to fiction. The lack of contextual literacy does not allow students to answer the question and thus the problem illustrates the importance of contextual literacy. After solving Unit 1 presented in this chapter the question becomes clearer and students can answer the question, saying that Michal was looking at a thermometer that measures temperature in degrees Celsius while Eran's thermometer showed the temperature in degrees Fahrenheit. Thus, one of the possible solutions could be that Michal wears a silk dress and a straw hat with a flower in Tel Aviv while Eran wears a warm coat and wool cap with a pompom in New York (see Figure 5.4). We can then ask the students (and the readers): "Can this happen on the same day?"

The Hat problem requires from solvers critical reasoning and contextual literacy (namely geographical and scientific literacy). While no mathematical competencies are applied to solving the Hat problem, it is used to motivate students when working on Unit 1. Moreover, mathematical literacy is activated to answer the following questions:

- If Eran was using Michal's thermometer, which temperature he would see?
- If Michal would use Eran's thermometer, which temperature it would depict?

Acknowledgment

The Math LIGHT program is developed with the generous support of the Julius and Edie Trump Family Foundation (Grant #358) and the Israel Science

Foundation (Grant #887/18). This publication reflects only the views of its author(s) and not necessarily those of the Foundations. Our sincere thanks go to the University of Haifa for its generous support of this study.

Notes

1 https://www.dictionary.com/browse/literacy
2 Math-LIGHT team: Sariga Dinur, George Ghantous, Irina Gurevitch, Alexander Levin, Yael Lin, Maayana Miskin, Regina Ovodenko, Zvia Parnes, and Shula Weisman.
3 Picture 1 – iStock-1344610065 –is downloaded from the website www.istockphoto.com – under the permission of the University of Haifa.
4 Picture 2.1 – iStock-1263557867 – is downloaded from the website www.istockphoto.com under the permission of the University of Haifa.
5 Picture 2.2 is produced by George Ghantous.
6 Pictures 2.3 and 2.4 are produced by Roza Leikin.
7 Picture 3.1 borrowed from https://www.vangoghmuseum.nl/en/collection/s0028V1962
8 Picture 3.2 is borrowed from https://artsandculture.google.com/asset/pont-neuf-paris/WAFRh_vU9ZvTPg?hl=en-GB
9 Target Problem 3 in the Unit "Perspective" is introduced to the students after they study construction of objects according to perspective laws.

References

Borromeo Ferri, R. (2006). Theoretical and empirical differentiations of phases in the modelling process. *ZDM – Mathematics Education, 38*(2), 86–95. https://doi.org/10.1007/BF02655883

Borromeo Ferri, R., & Blum, W. (2014). Barriers and motivations of primary teachers for implementing modelling in mathematics lessons. In B. Ubuz, C. Haser, & M. A. Mariotti (Eds.), *Proceedings of the 8th congress of the European Society for Research in Mathematics Education* (pp. 1000–1009). CERME.

Collins, J. W., & O'Brien, N. P. (2011). *The Greenwood dictionary of education.* ABC-CLIO.

De Lange J. (1996). Using and Applying mathematics in education. In A. J. Bishop, K. Clements, C. Keitel, J. Kilpatrick, & C. Laborde (Eds.), *International handbook of mathematics education* (pp. 49–98). Springer. https://link.springer.com/chapter/10.1007/978-94-009-1465-0_3

Freudenthal, H. (1968). Why to teach mathematics so as to be useful. *Educational Studies in Mathematics, 1*(1), 3–8. https://www.jstor.org/stable/3481973

Geiger, V., Forgasz, H., & Goos, M. (2015). A critical orientation to numeracy across the curriculum. *ZDM – Mathematics Education, 47*(4), 611–624. https://doi.org/10.1007/s11858-014-0648-1

Goldin, G. A., & McClintock, C. E. (1979). *Task variables in mathematical problem solving.* Franklin Institute Press.

Goos, M., Geiger, V., & Dole, S. (2013). Designing rich numeracy tasks. In C. Margolinas (Ed.), *Proceedings of ICMI study 22: Task design in mathematics education* (pp. 591–599). https://hal.archives-ouvertes.fr/file/index/docid/837488/filename/ICMI_STudy_22_proceedings_2013-FINAL_V2.pdf

Goos, M., O'Meara, N., Patrick Johnson, P., Fitzmaurice, O., & Guerin, A. (2023). Mathematical modelling as a stimulus for curriculum and instructional reform in secondary school mathematics. In R. Leikin (Ed.), *Mathematical challenges for all* (pp. 51–70). Springer. https://doi.org/10.1007/978-3-031-18868-8_4

Julie, C. (2006). Mathematical literacy: Myths, further inclusions and exclusions. *Pythagoras, 64*, 62–69. https://hdl.handle.net/10520/EJC20870

Kaiser, G., & Sriraman, B. (2006). A global survey of international perspectives on modeling in mathematics education. *ZDM – Mathematics Education, 38*(3), 302–310. https://doi.org/10.1007/BF02652813

Kaiser, G., & Stender, P. (2013). Complex modelling problems in co-operative, self-directed learning environments. In G. A. Stillman, G. Kaiser, W. Blum, & J. P. Brown (Eds.), *Teaching mathematical modelling: Connecting to research and practice* (pp. 277–293). Springer. https://doi.org/10.1007/978-94-007-6540-5_23

Kaiser, G., et al. (2017). Professional competencies of (prospective) mathematics teachers – cognitive versus situated approaches. *Educational Studies in Mathematics, 94*, 161–182. https://doi.org/10.1007/s10649-016-9713-8

Kilpatrick, J. (2001). Understanding mathematical literacy: The contribution of research. *Educational Studies in Mathematics, 47*, 101–116.

Kulm, G. (1979). The classification of problem-solving research variables. In G. A. Goldin & C. E. McClintock (Eds.), *Task variables in mathematical problem solving* (pp. 1–22). Franklin Institute Press.

Leikin, R. (2019). Stepped tasks: Top-down structure of varying mathematical challenge. In P. Felmer, P. Liljedahl, & B. Koichu (Eds.), *Problem solving in mathematics instruction and teacher professional development* (pp. 167–184). Springer. https://doi.org/10.1007/978-3-030-29215-7_9

Leikin R., & Ovodenko, R. (2021). Stepped tasks for complex problem solving: Top-down-structured mathematical activity. *For the Learning of Mathematics, 41*(3), 30–35.

McKenna, M. C., & Robinson, R. D. (1990). Content literacy: A definition and implications. *Journal of Reading, 34*(3), 184–186. https://www.jstor.org/stable/40014518

Organisation for Economic Co-operation and Development (OECD). (2010). *PISA 2009 results: What students know and can do: Student performance in reading, mathematics and science* (Vol. 1). OECD. https://doi.org/10.1787/9789264091450-en

Organisation for Economic Co-operation and Development (OECD). (2022). *PISA 2022 mathematics framework.* https://pisa2022-maths.oecd.org/#Contexts

Pollak, H. O. (1968). On some of the problems of teaching applications of mathematics. *Educational Studies in Mathematics, 1*(1/2), 24–30. https://www.jstor.org/stable/3481977

Schoenfeld, A. (1992). Learning to think mathematically: Problem solving, metacognition, and sense-making in mathematics. In D. Grouws (Ed.), *Handbook for research on mathematics teaching and learning* (pp. 334–370). MacMillan. https://doi.org/10.1177/002205741619600202

Silver, E. A., & Zawodjewsky, J. S. (1997). *Benchmarks of students understanding* (BOSUN) *project.* Technical Guide. LRDC.

Street, B. (2003). What's "new" in new literacy studies? Critical approaches to literacy in theory and practice. *Current Issues in Comparative Education, 5*(2), 77–91. https://doi.org/10.52214/cice.v5i2.11369

Street, B. (2006). Autonomous and ideological models of literacy: Approaches from New Literacy Studies. *Media Anthropology Network, 17*, 1–15. https://www.easaonline.org/downloads/networks/media/09p.pdf

UK National Literacy Trust. (2017). *What is literacy?* https://literacytrust.org.uk/information/what-is-literacy/

Venezky, R. L., Wagner, D. A., & Ciliberti, B. S. (Eds.). (1990). *Toward defining literacy.* International Reading Association. https://repository.upenn.edu/literacyorg_chapters/10

Vorhölter, K., Kaiser, G., & Borromeo Ferri, R. (2014). Modelling in mathematics classroom instruction: An innovative approach for transforming mathematics education. In Y. Li, E. Silver, & S. Li (Eds.), *Transforming mathematics instruction* (pp. 21–36). Springer. https://doi.org/10.1007/978-3-319-04993-9_3

Wallendorf, M. (2001). Literally literacy. *Journal of Consumer Research, 27*(4), 505–511. https://doi.org/10.1086/319625

Webb, N. (1979). Content and context variables in problem tasks. In G. A. Goldin & C. E. McClintock (Eds.), *Task variables in mathematical problem solving* (pp. 69–102). The Franklin Institute Press.

PART 3

Mathematical Literacy in Real World Tasks and Experiences

CHAPTER 6

Supporting Teachers to Develop Mathematical Literacy through Contemporary Financial Contexts

Carly Sawatzki, Jill Brown and Tim Powers

Abstract

In this chapter, we share research insights generated through our work with Australian secondary school teachers as part of the Economics + Maths = Financial Capability project. As part of this educational design research project, a course for teachers of Year 7–10 humanities and mathematics (secondary school students, 12–16 years of age) was designed and tested. The course aimed to strengthen teachers' interdisciplinary knowledge of finance-related curriculum and content, as well as their awareness of contemporary financial contexts, so as to enhance their ability to develop young people's financial capability. While the teachers we worked with tended to lack confidence in their mathematical knowledge and skills to address numeracy learning outcomes across the curriculum, they were keen to develop their understanding of contemporary financial contexts, as they recognised these as effective levers to explore mathematics with students in their classes. Our analysis shows that the course helped teachers to see the interconnected nature of economic and mathematical knowledge as it is drawn upon and applied in the real world, and opened new and powerful possibilities for their practice.

Keywords

financial literacy – financial capability – economics education – realistic mathematics education – teacher professional learning

1 Background

In Australia, the development of active and informed citizens is identified as a key goal of schooling (Australian Education Council, 2019). However, current approaches to financial education are failing to prepare young people for an increasingly complex economic and financial world. While recent Organisation

 | DOI:10.1163/9789004710962_006

for Economic Cooperation and Development (OECD) Programme for International Student Assessment (PISA) reports show that Australian students generally perform well on the financial literacy assessment, there has been little improvement over time and a high proportion of students score below the baseline level of proficiency (2015: 20%, 2018: 16%) (Thomson & de Bortoli, 2017; Thomson et al., 2020). Those who achieve below Level 2 have limited skills and are not yet able to apply their knowledge to real-world financial situations (Thomson et al., 2020). Hence the question, "How can we educate a financially capable citizenry?" has the attention of governments, policymakers, curriculum writers and researchers. Since the 2008 global financial crisis, efforts to enhance financial education in Australian schools have focused on the development of policy and instructional materials. This work has assumed that teachers want lesson plans and teaching resources and led to a proliferation of online content. Little attention has been paid to engaging teachers' intellect and imagination, and there remains significant potential to develop the teaching workforce with the essential knowledge for advancing young people's financial capability.

Through an educational design research project guided by theories of teacher professional knowledge and identity, we sought to gain a deeper understanding of how best to develop and support teachers as financial educators. The project was titled Economics + Maths = Financial Capability and involved an educational intervention (course) targeting teachers of Year 7–10 humanities and mathematics (secondary school students, 12–16 years of age) that aimed to strengthen teachers' interdisciplinary knowledge of finance-related curriculum, content, and pedagogy. Shulman's (1986, 1987) framework for understanding teachers' professional knowledge was used to analyse data collected via surveys and interviews with course participants, who were mainly teachers of mathematics working outside their main area of expertise ("out-of-field teachers" – Ingersoll, 2019) and teachers of humanities and social sciences working to recognise and exploit the numeracy demands that exist within this area of the curriculum ("non-specialist teachers of mathematics" – Goos et al., 2019).

2 Defining Key Terms

Various terms with nuanced meanings are used to describe the knowledge, skills and practices that financial education should aim to develop. These are: mathematical literacy (referred to as numeracy in Australia and throughout this chapter); financial literacy; financial numeracy; and financial capability. To understand the complex, messy space within which our work resides, it is important to distinguish between these terms and understand how they are applied in education discourses.

For the purpose of testing 15-year-olds' capacity to apply their knowledge and skills to real world problems, the OECD PISA defines *mathematical literacy* as:

> an individual's capacity to formulate, employ and interpret mathematics in a variety of contexts. It includes reasoning mathematically and using mathematical concepts, procedures, facts and tools to describe, explain and predict phenomena. It assists individuals to recognise the role that mathematics plays in the world and to make the well-founded judgments and decisions needed by constructive, engaged and reflective citizens. (OECD, 2017a, p. 67)

and *financial literacy* as:

> knowledge and understanding of financial concepts and risks, and the skills, motivation and confidence to apply such knowledge and understanding in order to make effective decisions across a range of financial contexts, to improve the financial well-being of individuals and society, and to enable participation in economic life. (OECD, 2017a, p. 87)

Consecutive OECD PISA reports reveal a strong correlation between mathematical and financial literacies (2015: 0.74, 2018: 0.87) (Thomson et al., 2020). However, the above definition of financial literacy has been critiqued as inadequate, as it implies a myopic focus on personal money management (earning, budgeting and saving) in a way that promotes individualism. Scholars have raised concerns that defining financial literacy in this way reflects a desire to socialise for economic rationalism and compliance (Retzmann & Seeber, 2016). The individual is cast as ultimately responsible for their own financial reality and education, with discourses leaning towards blaming poor socioeconomic outcomes on a lack of knowledge (Blue & Pinto, 2022; Davies, 2015).

While rooted in mathematics, numeracy aspires to mobilise the knowledge and skills of mathematics in ways that transcend the goals of mathematics (Liljedahl, 2015). As such, the term *financial numeracy*, which intends to capture the implicit mathematical demands associated with financial problems and decisions, is gaining traction among mathematics education researchers (Gal et al., 2020). Savard and Cavalcante (2021) define financial numeracy as consisting of contextual, conceptual and systemic dimensions. The contextual dimension is the staging of a mathematics task in a financial context. The conceptual dimension refers to the teaching and learning of financial concepts which require mathematical activity. The systemic dimension is the unpacking of financial concepts in relation to their social implications. Defining financial numeracy in terms of these dimensions aligns with models that depict the interplay between the mathematical and real worlds (comprehensively described by Stacey & Turner, 2015), including the 21C numeracy model (Goos et al., 2014) and mathematical modelling cycles (such as Stillman & Brown, 2021).

The 21C numeracy model identifies four key dimensions (contexts; mathematical knowledge and skills; tools; and dispositions) which are activated through a critical orientation. While initially conceived as a way to support teachers' planning for, and reflection on, teaching and learning numeracy, the model has also been used to guide the design of interdisciplinary learning tasks (e.g., Sawatzki, 2017; Sawatzki & Goos, 2018) and educate teachers (Goos et al., 2019).

Mathematical modelling cycles have been used to describe students' financial problem-solving and decision-making within primary and secondary school mathematics lessons (Sawatzki & Sullivan, 2018; Sawatzki et al., 2019, 2020). There are rich social and economic phenomena that mathematics can help to better understand, including employment patterns, wage gaps, inflation and interest rates, and fluctuations in investment prices. Stillman et al. (2020) argue that the mathematical practices used in the real world should be the practices valued and developed in school classrooms. Savard and Cavalcante (2021) argue that to truly understand important financial concepts (such as credit, investing, gambling) students must develop essential mathematical practices such as estimating, modelling, and predicting. For example, students can be taught to keep track of money by estimating costs and monitoring the income available to cover these. They can be taught to develop financial models that can inform more complex financial decisions, including about loans and investments. They can be taught how to draw on these financial models to predict such things as future risks, costs and rewards. Particularly in mathematics classrooms, lessons about finance should encourage students to develop models and then compare these to 'expert' models such as finance reports in the media.

In our work with Australian policymakers, curriculum writers and teachers (the audiences we hope our thinking will influence), we adopt the term *financial capability*. The reasons for this are practical and philosophical. Australian policy narratives have gradually shifted in the use of terms, with financial literacy being replaced by financial capability (we note that the term financial capability has always been used in New Zealand). Australia has a National

Financial Capability Strategy (Australian Government, 2022), previously overseen by the Australian Securities and Investments Commission (the financial regulator) and now by the Australian Government Treasury (the ministerial department responsible for providing the government with economic analysis and policy advice). The term financial capability encompasses the language, literacy and numeracy demands of financial problems and decisions in the real world. It also implies that financial understandings, choices and behaviours reflect social and cultural values and practices related to money (see Finley, 2021; Hunter & Sawatzki, 2019). Further, the term financial capability places stronger emphasis on preparing individuals to critique the role of governments and organisations within economies and financial systems (Sherraden & Ansong, 2016; Swalwell, 2021). We have found that the term financial capability resonates with Australian teachers, who are required to recognise and exploit the numeracy demands that exist across the Australian Curriculum.

3 Financial Education in Australia

The Australian Curriculum (AC) frames formal opportunities for students to learn about finance within humanities and social sciences (AC:HaSS, where economics and business is taught and learned) and mathematics (AC:M). This framing reflects the importance of economic knowledge and skills (Amagir et al., 2018; Retzmann & Seeber, 2016) and the relationship between economic, mathematical and financial literacies (Thomson et al., 2020). Additionally, numeracy is positioned as a general capability. All teachers are encouraged to help students to recognise the interconnected nature of knowledge as it is drawn upon and applied in the real world and give them opportunities to transfer and use their mathematical knowledge and skills outside the mathematics classroom. However, teachers have not been supported with detailed guidance to design learning tasks and sequences that embed numeracy across the curriculum, or to make decisions about pedagogies that support numeracy learning (Goos et al., 2013). So while the Australian Curriculum would seem to provide robust framing for financial education at school, there is a persistent question mark over whether the curriculum is achieving its intended goal. Schools and teachers are largely left to define and interpret what it means to be financially capable and how financial education at school can prepare financially capable young people.

Consistent with the above curriculum framing, Australian schools typically assign responsibility for financial education to teachers of HaSS and mathematics. In Australia, secondary school teachers receive specialised preparation that qualifies them to teach two disciplines for students in Years 7–12 (12–17-year-olds). Secondary school teaching allocations tend to be organised by disciplinary expertise. This results in students learning about finance in economics and/or mathematics lessons. Interdisciplinarity may be evident when teachers embed numeracy tasks in economic lessons and explore economics and finance in mathematics lessons. Sawatzki et al. (2017) found that secondary school commerce teachers in Australia were likely to be teaching mathematical knowledge and skills, including how to calculate cost per unit; how to read and interpret financial information presented in tables, charts and graphs; how to calculate simple and compound interest; and how to make sense of a payslip, including calculating taxation and superannuation payments. However, they struggled to imagine opportunities for students to explore economic, mathematical, and financial concepts in more contemporary real-world contexts. In contrast, Björklund (2019) found that secondary school social studies teachers in Sweden varied in their level of comfort with financial questions they deemed to be mathematical, commenting on their students' (and their own) uncertainty and struggles with numbers and calculations. Sawatzki et al. (2017) argued that a lack of knowledge and confidence likely contributes to teachers taking a somewhat simplistic focus on dry finance topics for personal money management, including budgeting, understanding and managing debt, and creating personal wealth through investing. This conventional approach has been critiqued for conveying a promise of sorts that encourages young people to believe that living within their means will not only safeguard against financial trouble, but create a trajectory towards personal wealth through investing (Pinto & Blue, 2016, 2017).

Since formal education in finance-related content and pedagogy is not an essential aspect of initial teacher education (e.g., Sweden: Björklund & Sandahl, 2020; Australia: Blue et al., 2014), education researchers and teacher educators argue a need for professional learning that prepares teachers to imagine and design more ambitious financial education that goes beyond personal money management (see Amagir et al., 2018; Björklund, 2019; Björklund & Sandahl, 2020). Yet in Australia, investments in teaching resources and online content have been prioritised over professional learning initiatives. Despite this, it is our experience working with hundreds of preservice and in-service teachers each year, that teachers believe that financial education is important and are motivated to hone their knowledge and skills in this area. Primary school teachers involved in Sawatzki and Sullivan's (2017) study distinguished between being financially literate themselves and having the curriculum, content and pedagogical expertise to develop students' financial literacy. While more than 75% of the teacher participants in that study agreed or strongly agreed that they were financially literate, only around half indicated being confident to develop students' financial literacy. Confidence grows with access to quality professional learning (Graven, 2004). Previous classroom research has demonstrated that interdisciplinary learning tasks that are relevant and useful to adolescents can be used to generate powerful teacher learning about finance-related curriculum, content and pedagogy (Hunter & Sawatzki, 2019; Sawatzki & Sullivan, 2017) as well as powerful student learning about economic, mathematical and financial concepts (Sawatzki, 2017; Sawatzki & Goos, 2018). In short, teachers and students benefit from opportunities to learn through contemporary financial contexts where curricular synergies exist (such as in commerce).

While the Australian Curriculum provides adequate framing to encourage schools to adopt disciplinary and interdisciplinary approaches to teaching and learning about finance, it appears rare to see teachers from different disciplines working together to design and teach financial education programs. Finance is typically viewed as a niche curriculum that only some teachers have the knowledge and confidence to enact. The myopic public view of what it means to be financially capable, and a lack of teacher education is perpetuating this situation. Yet we believe that it is possible to challenge and change this status quo and support teacher growth through professional learning where robust interdisciplinary learning tasks and intentional pedagogies are modelled and discussed (Liljedahl, 2015). This belief guided the research design.

4 The Research Design

4.1 *Theoretical Guidance*

The problem of being required to teach content outside one's comfort zone is not new. Shulman (1986, 1987) studied this and described teachers as having an intellectual biography, that is a "set of understandings, conceptions, and orientations that constitutes the source of their comprehension of the subjects they teach" (1986, p. 8). He established a framework for understanding teachers' professional knowledge that distinguishes between content knowledge, pedagogical content knowledge and curricular knowledge. Content knowledge refers to the amount and organization of subject matter knowledge in the mind of the teacher. Teachers must not only be capable of defining and explaining concepts to students, but also why this knowledge is worth knowing, and how it relates to other knowledge both within and outside the discipline. Pedagogical content knowledge goes beyond knowledge of subject matter to the dimension of subject matter knowledge for teaching. This type of knowledge is necessary to make instructional decisions about the most regularly taught topics in one's subject area, the most useful forms of representation of those ideas, and the most powerful analogies, illustrations, examples, explanations, and demonstrations that make the subject comprehensible and engaging to others. Curricular knowledge refers to the requirements for teaching particular subjects and topics at a given level, the variety of instructional materials available to resource those programs, and the criteria that guide teachers in selecting these. Curricular knowledge is both lateral and vertical. Lateral curriculum knowledge prepares teachers to relate the content of a given lesson to topics and concepts being explored in other classes. Vertical curriculum knowledge refers to familiarity with the topics and concepts that have been and will be taught in the same subject area during the preceding and later years in school. This framework acknowledges that teachers' knowledge and work are highly specialised.

Sachs (2005) argued that while teacher professional knowledge is of course important, it interacts with professional identity. Wenger (1998) explained that teachers are agents in shaping their own development, and professional learning provides a mechanism for moving from a current to a future professional identity. Since Australian teachers are required to demonstrate ongoing professional learning activity to maintain registration to teach, it makes sense that they will choose to pursue opportunities that align with their sense of self, what they believe is important in education, and what they think will be useful to the demands of their current teaching allocation. These factors explain why teachers do and don't opt in to professional learning on the topic of teaching about finance. Studies by Bennison (2015, 2022) provided useful theoretical and practical insight how to develop the capacity of non-specialist teachers of mathematics whilst shaping their identity as embedders-of-numeracy, in this case with a focus on contemporary financial contexts, across the curriculum.

4.2 *Methodology and Methods*

Educational design research methodology is highly regarded within mathematics education research for its strength in fostering productive learning ecologies and contributing to the development of local instructional theory (Gravemeijer & Cobb, 2006). Central to our ongoing program of research is the design and testing of interdisciplinary learning tasks for developing young people's financial capability. Through tasks that prompt a variety of solution strategies, our previous classroom research has been able to generate rich insights into the financial values and practices being learned within students' families and communities (see Sawatzki, 2017; Sawatzki & Goos, 2018) and the social, cultural and mathematical funds of knowledge that are drawn upon to solve financial problems (Bills et al., 2021; Hunter & Sawatzki, 2019). Since the COVID-19 pandemic prevented us from undertaking classroom research, we took our work with teachers online.

4.3 *The Economics + Maths = Financial Capability Course*

The educational intervention designed and tested was a course for teachers of Years 7–10 students (12–16 years of age). In designing the course, Carly and Jill considered what a financially capable young person might value, know, and do about money within their families and communities. For example, they would be literate and numerate, able to read, interpret and make decisions about economic and financial data in various forms, from the everyday (payslips) to the more complex (financial market reports). They would be organised in how they keep track of money and maintain financial records. They would be sceptical and critical thinkers, able to discern fact from fiction when engaging with such things as business marketing, government policy, and media reporting. They would be digitally astute in how they use financial technology ("fintech"), keeping their personal data safe, and creating a financial advantage. They would understand that people from different backgrounds view and experience money differently. They would possess an ethics of care in financial interactions, looking out for those who may be vulnerable and/or in financial trouble, directing them to seek trustworthy help from reputable sources without stigma. They would be future-focused and oriented to make spending and investment choices that reduce their social and environmental footprint to enhance not only their personal financial position, but the broader challenges humanity is facing. Importantly, they would be able to apply fundamental economic and mathematical knowledge and skills to the financial problems and decisions they face.

Based on this vision, the course aimed to highlight creative ways that teachers might read, interpret and contextualise the AC:HaSS and AC:M (curriculum and content knowledge, or *what to teach*) and model impactful pedagogical approaches (pedagogical content knowledge or *how to teach*). The course consisted of 6 × 90-minute seminars over 12 weeks. Topics were as follows:

1. Curriculum and pedagogy for developing financial capability;
2. Teenage tap n go;
3. All fun and games? A focus on music, entertainment, and gaming subscriptions;
4. The real cost of buying now and paying later;
5. Climate-conscious consumption: A focus on renewable energy; and
6. Climate-conscious consumption: A focus on new fashion trends.

The first seminar provided essential curriculum and research insights underpinning the series, including the 21C numeracy model (Goos, 2007; Goos et al., 2014) and the mathematical modelling cycle described by Stillman and Brown (2021). This foregrounded the fact that some of the interdisciplinary learning tasks that Carly and Jill design are numeracy tasks, and others are more demanding mathematical modelling tasks. Subsequent seminars explored contemporary financial contexts that young people might find interesting and useful to learn about. It is important to note that as the financial landscape is dynamically changing, a different selection of topics would likely be made today (only 18 months later) and/or initial learning materials would need to be updated.

Interdisciplinary learning tasks were central to the course design. Carly and Jill worked together to design tasks that would exploit key connections between HaSS and mathematics, as well as the Australian Curriculum's general capabilities. The tasks adhered to the design principles described by Galbraith et al. (2010) and intended to encourage a more holistic and modern approach to teaching about finance. Two illustrative examples are presented below.

4.3.1 Example 1: Best Buys with a Twist

The AC:M specifies that Year 7 students (i.e., 12 years of age) investigate and calculate 'best buys', with and without digital technologies (Mathematics, ACMNA174).

Typical textbook problems prompt students to discover that buying in bulk tends to lead to a reduction in the cost per unit. For example, a 5 kg bag of rice priced at $5 provides better value than a 1 kg bag of rice priced at $1.20.

We created a *Best buys with a twist* investigation that addressed the AC:HaSS requirement that Year 7 students consider the ways consumers and producers interact and respond to each other in the market (HaSS: Economics & Business, ACHEK017).

The question was: "Is it possible to reduce the use of single-use plastics within my home and save money by doing so?"

The open nature of the question is deliberate. The idea is that students choose to investigate something relevant to their family, do some research and run some calculations. Students would use mathematics to convince their teacher and peers that changing buying behaviour (or not) is worthwhile. This task empowers students to engage in the mathematical practices of mathematising, problem-solving and decision-making. It encourages students to prepare a cost-benefit analysis that compares products in plastic packaging with more sustainable alternatives. The cost per unit and/or cost per use can be calculated – but of greater interest will be the importance students place on minimising the impact of household purchases on the environment. Note the potential to challenge what is meant by the term 'best buy' – is 'best' always cheapest?

4.3.2 Example 2: Real Cost of Buying Now and Paying Later

The AC:M specifies that Year 9 students learn to solve problems involving simple interest (Mathematics, ACMNA211).

Typical textbook problems ask students to apply the simple interest formula to scenarios. For example, $1,000 is invested for 2 years at 5% simple interest. What is the total interest earned?

The dynamic nature of financial innovation means that students need practice applying literacy, numeracy and critical thinking skills to make informed decisions about new and unfamiliar financial products and services, including fintech solutions. Buy now pay later services provide a topical example. This new payment solution is popular among young people, who are averse to banks and credit cards. Buy now pay later providers lend differently. They are not classified as credit providers under the National Consumer Credit Protection Act 2009 (National Credit Act) because they don't technically charge interest. This means that they do not need to hold an Australian credit licence or comply with responsible lending obligations. The cost structure of such services (described in Figure 6.1) resembles simple interest.

We created a *Real cost of buying now and paying later* investigation that addressed the AC:HaSS requirement that students learn about the nature of innovation and how and why businesses seek to create and maintain a competitive advantage in the market, including the global market (Economics & Business, ACHEK041).

The task outlined BNPL's terms and conditions of use in language that is accessible to Year 9 students below (Figure 6.1). The following scenario was then put forward: "Kim needs your advice. She used BNPL to buy a reconditioned PlayStation for $265. Determine all possible scenarios and amounts that Kim might pay, depending on whether her payments are made on time. For each scenario, what is the effective interest rate?"

The course benefits marketed to teachers included that it was research-informed and designed by education experts; that they could study online via Zoom with independent, live, and small-group activities together with colleagues from across Australia; and that they would gain access to a secure course website containing readings, videos, podcasts, and presenter slides with the interdisciplinary learning tasks. In these ways, the course design reflected internationally recognised characteristics of effective professional learning (Desimone, 2009; OECD, 2017b). Since previous studies reveal teacher change occurs through enactment and reflection involving aspects of the teacher's world (Hollingsworth & Clarke, 2017; Lomas, 2018), we created opportunities for teachers to think about their professional knowledge, their professional identity, and their classroom practice through preliminary readings and breakout room discussions. As education researchers and teacher educators with complementary expertise in economics and mathematics, we modelled the teaching of the interdisciplinary learning tasks through intentional pedagogies. We

BNPL is a buy now pay later service provider created by Australian entrepreneurs. BNPL allows customers to split the total cost of a purchase over four equal instalments. The first time you use BNPL, the first payment is deducted at the point of purchase, with three further payments fortnightly thereafter. Next time you use BNPL, you get your goods straight away, but the first payment is not deducted until a fortnight after your purchase.

Payments are expected to be made on or before the due date. BNPL's key marketing messages include "no fees when you pay on time" and "never pay interest". However, customers need to read the fine print to understand what happens if they are late in making a payment. If you miss a payment, you will be prevented from using BNPL for any new purchases until your payments are up-to-date. If you don't have enough money in your nominated debit or credit account to cover the automatic payment and your payment is declined, you have 24 hours to log into your account and pay before BNPL charges you a $10 late fee. The late fee varies depending on the cost of your purchase. Caps (maximum fees) were introduced in 2018, meaning that the total you can be charged in late fees for any one purchase is 25% of your original order.

Some scenarios are outlined below:
- If your purchase was less than $40, you can only be charged a maximum of one $10 late fee for that order.
- If your purchase was between $40 and $272, an additional $7 late fee will be charged if the payment remains unpaid seven days after the due date.
- If your purchase was more than $272, the maximum late fee is $68 (i.e., 4 × ($10 + $7)).

FIGURE 6.1 BNPL terms and conditions

asked teachers to consider how their professional learning might be applied in their school setting to strengthen existing programs and/or create new ones.

Two iterations of the course were delivered: one in 2021 and one in 2022. The second iteration included minor adjustments to the course design. For example, the second cohort of teachers received an email one week prior to each seminar, which explained the topic in focus and gave a media article to hook them into the learning ahead. We also initiated the practice of briefly showing teachers the learning materials and teaching resources that they could access via the course website at the end of each seminar. Due to very low uptake, instructional recordings were offered to the first cohort, but not the second.

4.4 *Course and Research Participants*

Forty-seven secondary school teachers from 35 Australian schools (15 government, 7 Catholic, 13 independent; 26 metropolitan, 8 regional, 1 remote)

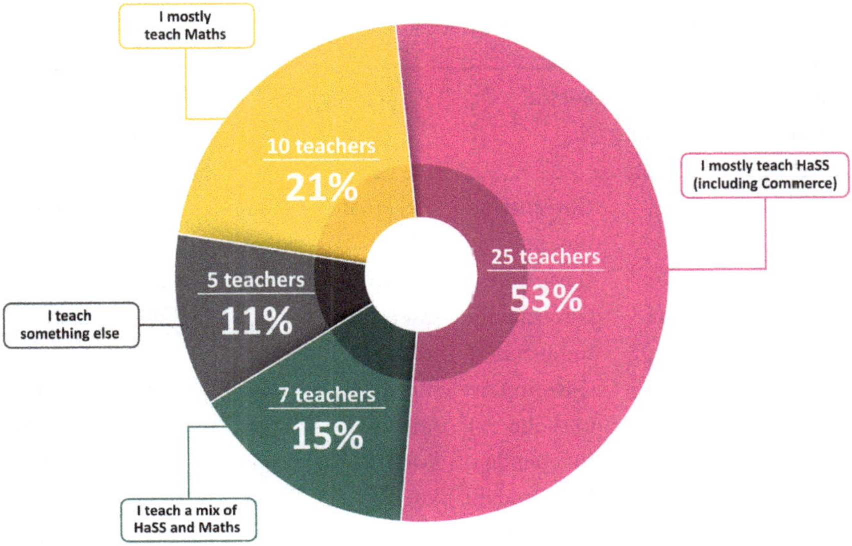

FIGURE 6.2 The teachers' disciplinary affiliations (n = 47)

completed the Economics + Maths = Financial Capability course. National reach was achieved, with teachers from all Australian states and territories participating.

The teachers were mainly teachers of mathematics working outside their main area of expertise ("out-of-field teachers" – Ingersoll, 2019) and teachers of HaSS working to recognise and exploit the numeracy demands that exist within this discipline ("non-specialist teachers of mathematics" - Goos et al., 2019). Figure 6.2 presents the disciplinary affiliations described by the 47 teachers.

The study involved two sources of data. Twenty-four teachers completed a course evaluation survey. Fourteen agreed to participate in a 30-minute semi-structured interview. Table 6.1 describes the professional characteristics of the teachers interviewed, together with information about their school sourced from the myschool website.[1] The Index of Community Socio-Educational Advantage (ICSEA), created by the Australian Curriculum, Assessment and Reporting Authority (ACARA), was used to understand the socioeconomic profile of each teacher's school community. An ICSEA value below the Australian average of 1000 is somewhat indicative of socio-educational disadvantage. Note that Cara, Keli, and Claudette were teaching in high-poverty schools.

4.5 *Data Analysis*

Interviews focused on teachers' perceptions of financial education at school (both generally and specific to teachers' school settings); the factors motivating them to enrol in the course; and their reactions to particular elements of

TABLE 6.1 Teachers interviewed and their school community

Teacher (pseudonym)	Years teaching	Sector	Regionality	School ICSEA	Indigenous students	LBOTE
Cara	5	Government	Regional	911	1%	90%
Stella	10	Independent	Metro	1,137	0%	9%
Keli	11	Government	Metro	938	2%	63%
Claudette	12	Government	Metro	955	1%	70%
Nerida	15	Independent	Metro	1,046	1%	40%
Rosa	17	Government	Regional	1,037	3%	6%
Alice	6	Catholic	Metro	1,158	1%	18%
Ruby	27	Independent	Metro	1,098	1%	9%
Anna	18	Independent	Metro	1,044	1%	33%
Rebekah[a]	23	Government	Metro	1,140	2%	34%
Owen[a]	5	Government	Metro	1,140	2%	34%
Carson	31	Government	Regional	941	12%	29%
Penelope	20	Government	Regional	999	3%	9%
Theresa	20	Independent	Metro	1,177	0%	52%

Note: ICSEA = Index of Community Socio-Educational Advantage; LBOTE = Language background other than English.
a Teachers at the same school.

the course design. Interviews were conducted online via Zoom by the third author after the third seminar and transcribed verbatim. The first and second authors drew on Shulman's (1986, 1987) framework for understanding teachers' professional knowledge to code and analyse survey data and interview transcripts. The interviews were analysed using constant comparative analytic procedures (Glasser & Strauss, 1967) to derive thematic categories. Becoming familiar with and analysing the data occurred over several meetings as we worked towards consensus in our sense-making. For the most part, the interview protocol posed questions that required teachers to reflect on their content knowledge, pedagogical content knowledge and curricular knowledge (i.e., Schulman's professional knowledge domains), and how these were being activated through their teaching, as well as developed through the course. This made the process of assigning initial codes aligned with the theoretical framing efficient. However, as we further examined what teachers were saying about each knowledge domain, we identified three thematic but interrelated categories: teachers' professional identity and disciplinary affiliation;

teachers' perceptions of mathematics and numeracy; and teachers' views on teaching through contemporary financial contexts. Within these categories, insights into teachers' knowledge and practice were evident, with dimensions of strength and insecurity revealed.

5 Findings

5.1 *Findings Drawn from Teacher Surveys*

Figure 6.3 presents teachers' survey responses reflecting on aspects of the course, including the seminar format, the learning materials, and the impact on their knowledge.

Teachers perceived the course to be useful in making an impact on their knowledge and confidence to teach about finance. Figure 6.4 presents teachers' reports of the impact of the program on their school and their students.

The vast majority of responding teachers indicated that the course strengthened their classroom teaching and prepared them to influence improvements to their school's existing programs. Around two thirds reported sharing course resources with colleagues within their school (sharing course resources with

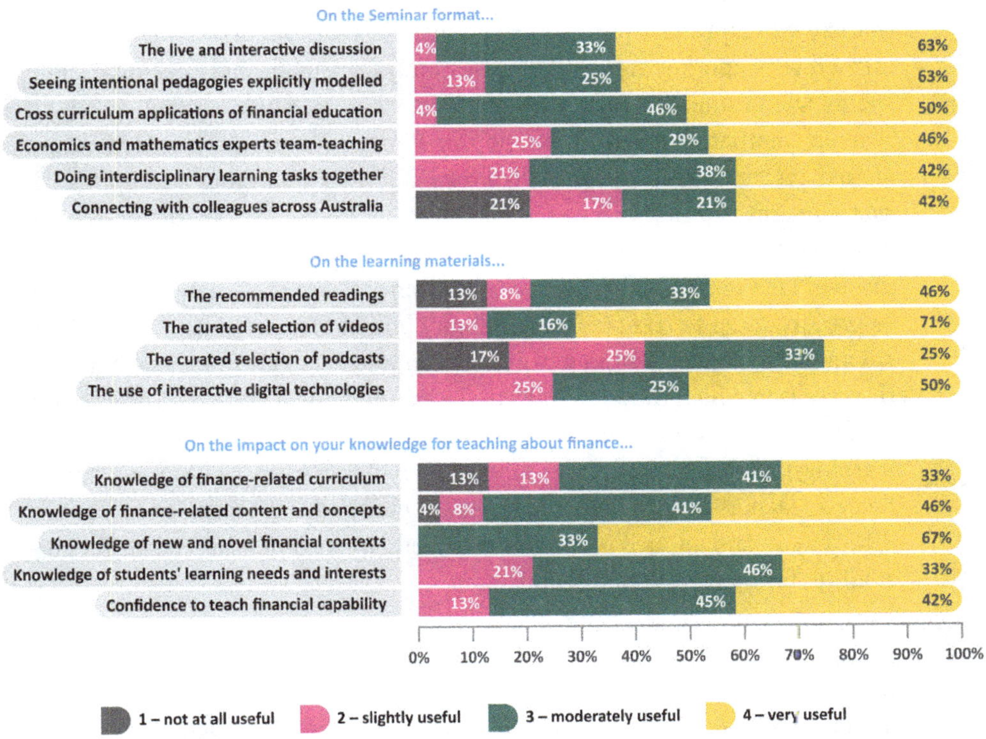

FIGURE 6.3 Teachers' ratings on the usefulness of aspects of the course (n = 24)

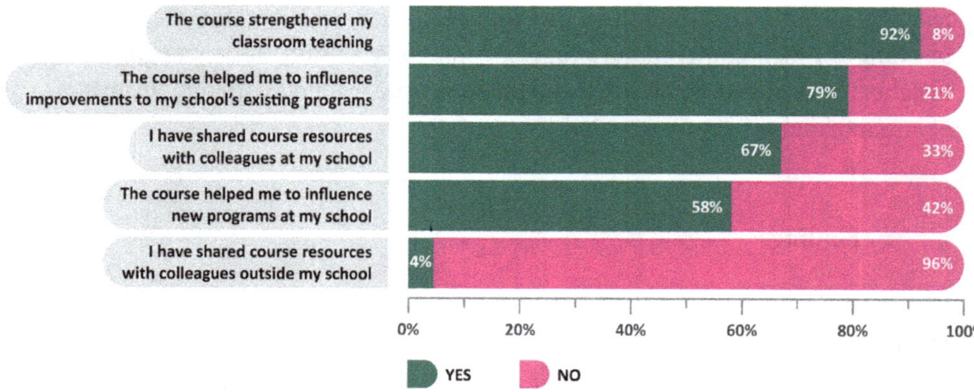

FIGURE 6.4 Indicators of teacher learning and agency as a result of the course (n = 24)

colleagues outside their school was rare). More than half reported that the course helped them to influence new programs at their school.

It is notable that these results were achieved in a relatively short period of time (i.e., 6 × 90-minute sessions over a 12-week period).

5.2 *Findings Drawn from Teacher Interviews*

The teachers we interviewed were mostly non-specialist teachers of mathematics, with a primary professional knowledge base and professional identity aligned with another area of the curriculum. They made clear personal statements which spoke to their subject affiliation as well as their feelings about teaching mathematics and/or developing student numeracy. The teachers also shared descriptive insights into their own and their colleagues' perceptions of mathematics and numeracy and how these are taught and learned at school. This suggested that the course was achieving its goal to disrupt teachers' traditional views of "What is mathematics?" and "What is numeracy?" as a way to prepare them to transcend disciplinary boundaries, as well as mathematical and real worlds. A source of frustration and fascination for teachers was the extent to which they saw links to the mathematical and real worlds being made in their schools. Our analysis shows that the use of interdisciplinary learning tasks involving contemporary financial contexts within the course helped teachers to recognise the interconnected nature of knowledge as it is drawn upon and applied in the real world and opened new and powerful possibilities for them as financial educators.

5.2.1 Teachers' Professional Identity and Disciplinary Affiliation

As they introduced themselves and their work, teachers identified as subject matter experts in the discipline they studied and typically taught. Out-of-field and non-specialist teachers of mathematics varied in their knowledge and confidence to engage with the numeracy demands associated with the interdisciplinary learning tasks, and expressed insecurity about this. For example, Cara worked in the finance industry prior to becoming a teacher of English, humanities and commerce. While this personal history meant she felt somewhat confident to teach students about finance, she was apprehensive to contribute to seminar discussions about the more challenging mathematical modelling tasks exploring compound interest that were designed for the course:

> I'm actually not a maths person. And I have to say, I really feel that when we discuss the maths in the [seminars] that we do. But I understand business, and I understand economics… I can't really speak to what's taught in the maths faculty at school.

Rebekah, another humanities teacher, expressed similar insecurities. Also reflecting on one of the mathematical modelling tasks exploring climate-conscious consumption (see Sawatzki et al., 2022), she explained:

> I have to admit I did not contribute [to the breakout room discussion] – it was over my head! Sorry! But I would still use the task in my teaching – in humanities classes – but slightly differently. I think it is a really important issue to discuss.

While lacking confidence in mathematics affected Cara and Rebekah's readiness to participate fully when mathematical workings were being generated and shared, they were among a number of teachers who reported that the course had prompted them to realise the mathematical and numeracy demands inherent in everyday financial problems and decisions. For example, Penelope, a commerce teacher, described that the interdisciplinary learning tasks had extended her knowledge and practice into mathematics and numeracy and opened new and powerful possibilities for her students' learning in this area:

> The way they teach about financial capability is interesting, because it links with numeracy. And I've never really considered it linking with numeracy. I'd only really thought about it from an economics or a business angle. But looking at it from a numeracy angle, there was so much cross curricular there, that it sparked my interest.

For Theresa, also a commerce teacher, the course affirmed the interdisciplinary nature of her existing knowledge and practice – an outcome that she did not expect. As she reflected on her learning, Theresa described the realisation that drawing on students' mathematical knowledge and skills to develop their financial capability was already a strength of her teaching:

> Actually, [the course] has been a little bit less mathsy than I was expecting. But I think maybe that's because I forget that what I'm doing in my classroom, there's already quite a lot of maths in it. See, I don't see things like a ratio or the downward slope or the demand curve as maths. I think that's economics. But I forget, it's actually maths too!

5.2.2 Teachers' Perceptions of Mathematics and Numeracy

A common issue raised by teachers in their interviews was a traditional view of mathematics and how it is taught and learned. Owen, an in-field Mathematics teacher, described feeling obligated to teach mathematics in a particular way:

> If you've had a look at the maths curriculum, the way we have to teach it is that roughly every two or three weeks, we've got to start teaching a new unit. And a lot of those units to my mind are so abstract that you lose [students]. Especially the Year 10s. By the time they've got to Year 10, they're not interested in maths anymore, which is really sad.

Owen saw contextualisation as the key to keeping students engaged with learning mathematics meaningfully and for longer, and gave excellent examples of investigations involving financial contexts that he had designed. His professional identity and agency were evident in his explanation, "I force it in there".

Keli, a commerce and [out-of-field] mathematics teacher who had also worked in the finance industry, identified that content at the intersection of humanities and mathematics could be explored across the curriculum, but expressed frustration that these connections were not being exploited at her school:

> So does a maths teacher who's teaching simple interest understand [they're] actually teaching financial [capability]? Or [do they] think [they're] teaching a calculation? If you don't give any of these calculations a context, the student is not financially [capable]. They just know how to calculate simple interest. That has no bearing on a decision later in life… For me, it's a frustration I've had, in my ten years of teaching, how little the students know coming into my Year 11 economics class…

> They cannot apply the maths they've learned to an economic situation. So, I'll give you a really, really simple example. In economics, we learn very quickly, you know, concepts like marginal utility, or what's another, elasticity. These are Year 7 maths calculations and the kids can't do them. So I have to re-teach the maths, because when the maths was taught to them, it didn't have a context… This is just a percentage change calculation, but there's no registering there of what I mean by that… And I've got some of the smartest kids in the school in my Year 11 economics class.

Stella, a teacher of Humanities and Commerce, also conveyed her impression that the mathematics teachers at her school lacked the knowledge and confidence to teach about finance in authentic and applied ways:

> Well, I think students just think it's boring. And it's just almost like a bit of maths. Do you know what I mean? So if you're not into maths, then students aren't going to find that really appealing.

Teachers attributed the issue to a reliance on textbooks for curricular direction, and the practice of setting numerous exercises of seemingly meaningless calculations. They described these default practices as contributing to students switching off from important learning opportunities. Claudette, a teacher of science and [out-of-field] mathematics, explained, "I feel like we're very limited to what's in the textbook, and we kind of follow what's in the textbook". Carson, a humanities teacher, shared the view that subject matter textbooks offer a narrow interpretation of the curriculum content and lack currency:

> I think where, where the biggest issue is at the moment, is keeping up with current trends and technologies. And I guess personally, that's where I have a difficulty with textbooks, because textbooks can become quickly outdated within, you know, 12 months. And it was interesting in the last session, looking at all the tap 'n' go type of things that are happening in this world. And I think as teachers, unless we actually use those sorts of products, we're pretty much in the dark. So it takes a lot of work for teachers to come up to speed on what those are, what the new technologies are.

Several teachers identified the challenge of knowing the expansiveness of the school curriculum and the work of colleagues in other disciplines. Rosa, a commerce teacher, expressed a level of insecurity about this, and conceded that developing her lateral curricular knowledge would likely enhance her capacity to teach about finance:

> Obviously, they're teaching financial mathematics over in the maths department. But there's not a lot of cross fertilisation going on between the departments. So it's really hard to know what they're teaching… I can't speak from a mathematical point of view, to be honest, I don't know their curriculum very well. I probably should, but I don't.

Claudette echoed this insight, and indicated wanting to work with colleagues with complementary knowledge to continue to develop her knowledge and practice beyond the course:

> When you teach a subject, you don't really see what other subjects do. So, you've got a very limited view about what actually goes on in the school… Yeah, I think there is kind of the view it's in somebody else's domain… So, I share a staff room with the maths faculty… So, we have lunch together… So, we sit there and discuss, "How can we do this better?"… I was actually surprised how many people from humanities… have been in the course that I've met in the breakout groups. And I'm thinking, well, maybe we should be working together more?

5.2.3 The Appeal and Power of Contextualised Teaching and Learning

The interdisciplinary learning tasks were often cited as helping the teachers to see new ways to contextualise their teaching and meet students' ever-changing needs and interests. For example, Claudette joined the course because she wanted to make her mathematics lessons as realistic and authentic as possible. She explained that the contemporary financial contexts that the course exposed her to had given her new ways to connect with and relate to students:

> I really like the examples they've been giving [in the course] where they've been using, you know, the currencies that exist in [Xbox and PlayStation] games. Because the kids are familiar with that.

Ruby, a commerce and [out-of-field] mathematics teacher, also reflected on the power of the approach being modelled through the course to engage students:

> I just love the way the course is touching on so many highly relevant things in kids' lives and connecting them to the curriculum. Now, the joke in maths is you always have questions like, 'Someone's got 63 watermelons…'. No one would ever have 63 watermelons! The questions that we're discussing in this course are just more relevant and related to the real-life experiences of kids.

6 Discussion

The Australian Curriculum assumes that all teachers will readily identify the numeracy demands that exist across the curriculum and support students to apply their mathematical knowledge and skills outside the mathematics classroom. However, teachers don't know what they don't know: they need access to expert guidance to develop their knowledge to connect complementary disciplines, and to connect mathematical and real worlds.

By their own admission, an over-reliance on textbooks as one interpretation of how the written curriculum might be enacted typically limits teachers' use of contemporary financial contexts. An explanation for this default practice is that while the knowledge teachers require to develop student financial capability is both extensive and complex, the education of teachers as financial educators is typically not prioritised. Hence teachers defer to textbooks for information and ideas. To excite teachers and students, and demonstrate the relevance and usefulness of mathematics, it is essential to connect mathematics with the real world. Contemporary financial contexts offer a way to do this and are typically well-received by teachers and students.

As we began to test our interdisciplinary learning tasks with teachers, we came to understand that the contemporary financial contexts carried economic and financial concepts and language that were new to some. Also, some teachers reported that the mathematical and numeracy demands of the tasks put their mathematical knowledge and skills to the test. Completing the tasks in breakout rooms and experiencing intentional pedagogies being modelled inspired the teachers while developing their knowledge and confidence to work in interdisciplinary ways. In effect, through the course, we potentialised financial education at school.

7 Concluding Remarks

The insights from teachers reported in this chapter explain why teaching and learning about finance is a sticky issue in education. Our findings confirm that teacher professional learning initiatives must be designed to develop teachers' knowledge across all three domains (curriculum knowledge, content knowledge, and pedagogical content knowledge). The Economics + Maths = Financial Capability course made explicit the finance-related aspects of the Australian Curriculum, and developed teachers' economic and mathematical content knowledge. Exposing teachers to contemporary financial contexts brought teachers closer to the lived financial activities and experiences of their students, thereby enhancing their pedagogical content knowledge. Strategies intended to build lateral and vertical curriculum knowledge gave teachers a new and more holistic view of the student learning experience and trajectory over the course of Years 7–10, and opened new possibilities for innovative, interdisciplinary teaching programs and practices within their schools. At the end of the course, the teachers reported gains in knowledge and confidence to develop young people's financial capability.

While the Economics + Maths = Financial Capability course provides a model for supporting teacher knowledge growth, we did encounter challenges. The course was offered at the height of the COVID-19 pandemic, around half of the teachers self-funded to attend, and most gave their personal time to engage with learning opportunities. These factors made recruitment highly challenging and contributed to absences and apologies. We also found that due to current teaching allocations and the scope and sequencing of the school curriculum, most teachers did not have immediate opportunities to experiment with the interdisciplinary learning tasks in their own classrooms, although plans for next semester and/or next year were often signalled.

We gathered teachers' views across a range of disciplines and demographics, but acknowledge that our study was affected by self-selection bias. All of the teachers who enrolled already identified themselves as financial educators and saw teaching about finance as an important aspect of their work. We did not manage to recruit teachers who were not already teaching about finance in a meaningful way.

Success stories that we encountered were driven by school leaders supporting interdisciplinary partners or teams to enrol and learn together. This confirms that school leaders and teachers are agentic and will drive the instructional and pedagogical changes they believe are important. School leaders and teachers who are reading this chapter and interested to do this might take initial steps by reflecting on these questions:

1. In what ways are students at your school already financially active?
2. What knowledge and skills (including economic and mathematical) do their current and future financial problems and decisions require?
3. What contemporary financial contexts are interesting and risky to them?
4. How might you easily embed teaching and learning about these financial contexts within existing learning programs?

To improve students' access to financial education at school, decision makers within education and curriculum authorities must incentivise teachers to see teaching about finance as a core part of their work and undertake professional learning on this topic.

Note

1 https://www.myschool.edu.au

References

Amagir, A., Groot, W., van den Brink, H. M., & Wilschut, A. (2018). A review of financial-literacy education programs for children and adolescents. *Citizenship, Social and Economics Education, 17*(1), 56–80.

Australian Education Council. (2019). *Alice Springs (Mparntwe) education declaration.* https://www.education.gov.au/alice-springs-mparntwe-education-declaration/resources/alice-springs-mparntwe-education-declaration

Australian Government. (2022). *National financial capability strategy.* https://files.moneysmart.gov.au/media/vyfbpg4x/national-financial-capability-strategy-2022.pdf

Bennison, A. (2015). Developing an analytic lens for investigating identity as an embedder-of-numeracy. *Mathematics Education Research Journal, 27*(1), 1–19. https://doi.org/10.1007/s13394-014-0129-4

Bennison, A. (2022). Using zone theory to understand teacher identity as an embedder-of-numeracy: An analytical framework, *Asia-Pacific Journal of Teacher Education, 50*(2), 171–186. https://doi.org/10.1080/1359866X.2020.1828821

Bills, T., Sawatzki, C., Hunter, J., & Miller, J. (2021). “Pākehā get more money than the other cultures”: Teaching Pāsifika students with and for a social justice orientation. In T. Lucey (Ed.), *Financialization, financial literacy and social education* (pp. 23–41). Routledge.

Björklund, M. (2019). Teaching financial literacy: Competence, context and strategies among Swedish Teachers. *Journal of Social Science Education 18*(2), 28–48. https://doi.org/10.4119/jsse-1426

Björklund, M., & Sandahl, J. (2020). Financial literacy as citizenship education – a viable prospect? *Journal of Social Science Education, 19*(3), 4–20. https://doi.org/10.4119/jsse-3230

Blue, L., Grootenboer, P., & Brimble, M. (2014). Financial literacy education in the curriculum: Making the grade or missing the mark? *International Review of Economics Education, 16*, 51–62. https://doi.org/10.1016/j.Iree.2014.07.005

Blue, L., & Pinto, L. (2022). Disrupting the alibi: Toward a postcolonial financial literacy and entrepreneurship ideal. In T. Lucey (Ed.), *Financialization, financial literacy and social education* (pp. 5–22). Routledge.

Davies, P. (2015). Towards a framework for financial literacy in the context of democracy. *Journal of Curriculum Studies, 47*(2), 300–316. https://doi.org/10.1080/00220272.2014.934717

Desimone, L. M. (2009). Improving impact studies of teachers' professional development: Toward better conceptualizations and measures. *Educational Researcher, 38*, 181–199. https://doi.org/10.3102/0013189X08331140

Finley, S. Y. (2021). Financial literacy, financial liberation: Toward a critical race approach to financial education. In T. Lucey (Ed.), *Financialization, financial literacy and social education* (pp. 113–127). Routledge.

Gal, I., Grotlüschen, A., Tout, D., & Kaiser, G. (2020). Numeracy, adult education, and vulnerable adults: A critical view of a neglected field. *ZDM Mathematics Education, 52*(3), 377–394. https://doi.org/10.1007/s11858-020-01155-9

Galbraith, P., Stillman, G., & Brown, J. (2010). Turning ideas into modelling problems. In R. Lesh, P. Galbraith, C. Haines, & A. Hurford (Eds.), *Modeling students' mathematical modeling competencies* (pp. 133–144). Springer.

Glasser, B. G., & Strauss, A. L. (1967). *The discovery of grounded theory: Strategies for qualitative research*. Aldine.

Goos, M., Bennison, A., Quirke, S., O'Meara, N., & Vale, C. (2019). Developing professional knowledge and identities of non-specialist teachers of mathematics. In D. Potari & O. Chapman (Eds.), *International handbook of mathematics teacher education* (Vol. 1). Oxford.

Goos, M., Geiger, V., & Dole, S. (2013). Designing rich numeracy tasks. In C. Margolinas (Ed.), *Task design in mathematics education. Proceedings of ICMI study 22* (Vol. 1). Oxford.

Goos, M., Geiger, V., & Dole, S. (2014). Transforming professional practice in numeracy teaching. In Y. Li, E. Silver, & S. Li (Eds.), *Transforming mathematics instruction. Advances in mathematics education*. Springer. https://doi.org/10.1007/978-3-319-04993-9_6

Goos, M., Stillman, G., & Colleen, V. (2007). *Teaching secondary school mathematics: Research and practice for the 21st century*. Allen & Unwin.

Graven, M. (2004). Investigating mathematics teacher learning within an in-service community of practice: The centrality of confidence. *Educational Studies in Mathematics, 57*(2), 177–211. https://doi.org/10.1023/B:EDUC.0000049277.40453.4b

Gravemeijer, K., & Cobb, P. (2006). Design research from a learning design perspective. In J. van den Akker, K. Gravemeijer, S. McKenney, & N. Nieveen (Eds.), *Educational design research*. Taylor & Francis.

Hollingsworth, H., & Clarke, D. (2017). Video as a tool of focusing teacher self reflection: Supporting and provoking teacher learning. *Journal of Mathematics Teacher Education, 20*(5), 457–475. https://doi.org/10.1007/s10857-017-9380-4

Hunter, J., & Sawatzki, C. (2019). Discovering diverse students' funds of knowledge related to finance: Pāsifika students in New Zealand. *Mathematics Education Research Journal, 31*(4), 419–439. https://doi.org/10.1007/s13394-019-00259-0

Ingersoll, R. M. (2019). Measuring out-of-field teaching. In L. Hobbs & G. Törner (Eds.), *Examining the phenomenon of "teaching out-of-field": International perspectives on teaching as a non-specialist* (pp. 21–52). Springer.

Liljedahl, P. (2015). Numeracy task design: A case of changing mathematics teaching practice. *ZDM Mathematics Education, 47*(4), 625–637. https://doi.org/10.1007/s11858-015-0703-6

Lomas, L. (2018). Proposed structural refinements to the interconnected model of teacher professional growth. In J. Hunter, P. Perger, & L. Darragh (Eds.), *Proceedings of the 41st annual conference of the mathematics education research group of Australasia* (pp. 495–502). MERGA.

Organisation for Economic Co-operation and Development (OECD). (2017a). *PISA 2015 assessment and analytical framework: Science, reading, mathematics, financial literacy and collaborative problem solving* (Rev. ed.). OECD. https://doi.org/10.1787/9789264281820-en

Organisation for Economic Cooperation and Development. (2017b). *Teaching in focus: How can professional development enhance teachers' classroom practices?* (OECD Education Working Paper No. 16). OECD Publishing. https://doi.org/10.1787/2745d679-en

Pinto, L. E., & Blue, L. E. (2016). Pushing the entrepreneurial prodigy: Canadian aboriginal entrepreneurship education initiatives. *Critical Studies in Education, 57*(3), 358–375. https://doi.org/10.1080/17508487.2015.1096291

Pinto, L. E., & Blue, L. E. (2017). Aboriginal entrepreneurship financing in Canada: Walking the fine line between self-determination and colonization. *Journal of Entrepreneurship in Emerging Economies, 9*(1), 2–16. https://doi.org/10.1108/JEEE-10-2015-0059

Retzmann, T., & Seeber, G. (2016). Financial education in general education schools: A competence model. In C. Aprea, E. Wuttke, K. Breuer, N. K. Koh, P. Davies, B. Greimel-Fuhrmann, & J. S. Lopus (Eds.), *International handbook of financial literacy* (1st ed., p. 9). Springer.

Sachs, J. (2005). Teacher education and the development of professional identity: Learning to be a teacher. In P. M. Denicolo & M. Kompf (Eds.), *Connecting policy and practice: Challenges for teaching and learning in schools and universities* (pp. 5–21). Routledge, Taylor and Francis Group. https://doi.org/10.4324/9780203012529

Savard A., & Cavalcante A. (2021). Financial numeracy as part of mathematics education. In: A. Savard & A. Cavalcante (Eds.), *Financial numeracy in mathematics education. Mathematics education in the digital era* (Vol. 15). Springer. https://doi.org/10.1007/978-3-030-73588-3_2

Sawatzki, C. (2017). Lessons in financial literacy task design: Authentic, imaginable, useful. *Mathematics Education Research Journal, 29*(1), 25–43. https://doi.org/10.1007/s13394-016-0184-0

Sawatzki, C., Brown, J., & Zmood, S. (2022). Using mathematics to drive future-focused financial capability. *The Australian Mathematics Education Journal, 4*(1), 25–30.
Sawatzki, C., Downton, A., & Cheeseman, J. (2019). Stimulating proportional reasoning through questions of finance and fairness. *Mathematics Education Research Journal, 31*(4), 465–484. https://doi.org/10.1007/s13394-019-00262-5
Sawatzki, C., & Goos, M. (2018). Cost, price and profit: What influences student's decisions about fundraising? *Mathematics Education Research Journal, 30*(4), 525–544. https://doi.org/10.1007/s13394-018-0241-y
Sawatzki, C. M., & Sullivan, P. A. (2017). Teachers' perceptions of financial literacy and the implications for professional learning. *Australian Journal of Teacher Education, 42*(5), 51–65. https://doi.org/10.14221/ajte.2017v42n5.4
Sawatzki, C. M., & Sullivan, P. A. (2018). Shopping for shoes: Teaching students to apply and interpret mathematics in the real world. *International Journal of Science and Mathematics Education, 16*, 1355–1373. https://doi.org/10.1007/s10763-017-9833-3
Sawatzki, C., Zmood, S., & Davidson, A. (2020). Using financial modelling to decide, "Should I tap into my superannuation?" *The Australian Mathematics Education Journal, 2*(2), 22–27.
Sawatzki, C., Zmood, S., Forsyth, A., & Downton, A. (2017). *Exploring secondary commerce teachers' opportunities and readiness to teach consumer, economic and financial literacy.* Research Report for the Victorian Commercial Teachers Association. https://www.vcta.asn.au/documents/item/3244
Sherraden, M. S., & Ansong, D. (2016). Financial literacy to financial capability: Building financial stability and security. In C. Aprea, E. Wuttke, K. Breuer, N. K. Koh, P. Davies, B. Greimel-Fuhrmann, & J. S. Lopus (Eds.), *International handbook of financial literacy* (1st ed., p. 83). Springer.
Shulman, L. (1986). Those who understand: Knowledge growth in teaching. *Educational Researcher, 15*(2), 4–14. https://doi.org/10.3102/0013189X015002004
Shulman, L. (1987). Knowledge and teaching: Foundations of the new reform. *Harvard Educational Review, 57*(1), 1–22. http://dx.doi.org/10.17763/haer.57.1.j463w79r56455411
Stacey, K., & Turner, R. (2015). The evolution and key concepts of the PISA mathematics frameworks. In K. Stacey & R. Turner (Eds.), *Assessing mathematical literacy: The PISA experience*. Springer.
Stillman, G. A., & Brown, J. P. (2021). Modelling the phenomenon versus modelling the data set. *Mathematical Thinking and Learning*, 1–26. https://doi.org/10.1080/10986065.2021.2013144
Stillman, G., Brown, J., & Czocher, J. (2020). Yes, mathematicians do X so students should do X, but it's not the X you think! *ZDM Mathematics Education, 52*(5), 1211–1222. https://doi.org/10.1007/s11858-020-01183-5
Swalwell, K. (2021). Noticing and questioning capitalism with elementary students. In T. Lucey (Ed.), *Financialization, financial literacy and social education* (pp. 42–60). Routledge.

Thomson, S., & De Bortoli, L. (2017). *PISA 2015: Financial literacy in Australia.* Australian Council for Educational Research. https://www.voced.edu.au/content/ngv%3A77902

Thomson, S., De Bortoli, L., Underwood, C., & Schmid, M. (2020). *PISA 2018: Financial literacy in Australia.* Australian Council for Educational Research. https://research.acer.edu.au/ozpisa/48

Wenger, E. (1998). *Communities of practice: Learning, meaning, and identity.* Cambridge University Wenger-Traynerity Press.

CHAPTER 7

Mathematical Literacy in the Context of Statistics

A Theoretical Discussion and an Example from a Research Project

Jonas Bergman Ärlebäck, Peter Frejd, Robert Schaller and Patrick Bergman Skiba

Abstract

This chapter discusses mathematical literacy in the context of the teaching and learning of statistics at the secondary level (Grades 7–9). This includes exemplifying how tasks and activities in statistics can be designed and implemented to support students in developing mathematical literacy using the models and modelling theoretical perspective and model development sequences. The chapter discusses literature on statistical literacy and related notions in the field of statistics education and links these to mathematical literacy more generally. Connections are made and highlighted between mathematical literacy and tenets of the models and modelling perspective as well as model development sequences. These connections are then illustrated as they come to the fore in an example of an activity from a research project focusing on the teaching and learning of statistics, which sheds light on the students' opportunities to develop their mathematical literacy.

Keywords

models and modelling perspective – model development sequences – mathematical literacy – statistical literacy

1 Introduction

The SARS-CoV-2 virus that swept over the world starting in late 2019 is the single event in recent times that has made statistical data and statistical arguments a part of people's everyday (Watson & Callingham, 2020) and professional lives (UNEP, 2020a, 2020b, 2021). It has also affected what it means to be a well-informed and responsible citizen (Heyd-Metzuyanim et al., 2021; Aguilar & Castaneda, 2021). Indeed, in public discourses surrounding the SARS-CoV-2

 | DOI:10.1163/9789004710962_007

pandemic, multiple strategies based on relations between data, data-based arguments and inferred relationships have been used in motivating, reinforcing, and maintaining public-health related rules and regulations controlling people's lives (Jablonka & Bergsten, 2021). Altogether, this has resulted in an increased awareness about the importance and urgent need to enhance the readiness among citizens of the world in general, and students in particular, to meet, interpret, evaluate, and make sense of real-world data. However, these insights and arguments are not new. Steen (1999) discussed what he called a "data-drenched society" and pointed out the increasing need for students to have the knowledge and adequate skillset to critically use and understand the various quantitative methods that come with the increasing amount and increased accessibility of data available due to the rapid technological development. Data per se does not, however, directly convey any information; rather, one needs models to cope with, interpret and make sense of data in private, societal, and professional life (Manyika et al., 2011). In such sense-making contexts, statistics, statistical reasoning and statistical models are invaluable tools for all students to become proficient in understanding a range of statistical topics and to learn the forms of reasoning that are needed to interpret the data (Franklin et al., 2007; OECD, 2013).

In this chapter, we discuss frameworks of *statistical literacy* (SL) to address the need for students to develop competencies to handle the data they encounter. The intention of the chapter is both to theoretically elaborate on the notion and content of SL using a *mathematical literacy* (ML) framework, and to illustrate how activities in statistics can be designed and implemented to support students in developing ML and SL using the *models and modelling theoretical perspective on teaching and learning* (Lesh & Doerr, 2003). We do this by first providing a reference framework of ML before briefly elaborating on statistics as a discipline and what it means to engage in statistical thinking and statistical problem solving. Then, we discuss SL and continue to point out connections between ML and SL, and tenets of the *models and modelling perspective* and *model development sequences* as a framework for designing and implementing the teaching and learning of mathematics. These connections are then illustrated using an example of an activity from a research project focusing on the teaching and learning of statistics to highlight the students' opportunities to develop their ML and SL based on this framework for design and implementation.

1.1 *Mathematical Literacy – A Frame of Reference*

The meaning and use of the notion of ML, and similar related notions[1], varies in the mathematics education literature (Geiger et al., 2015a; Jablonka, 2003;

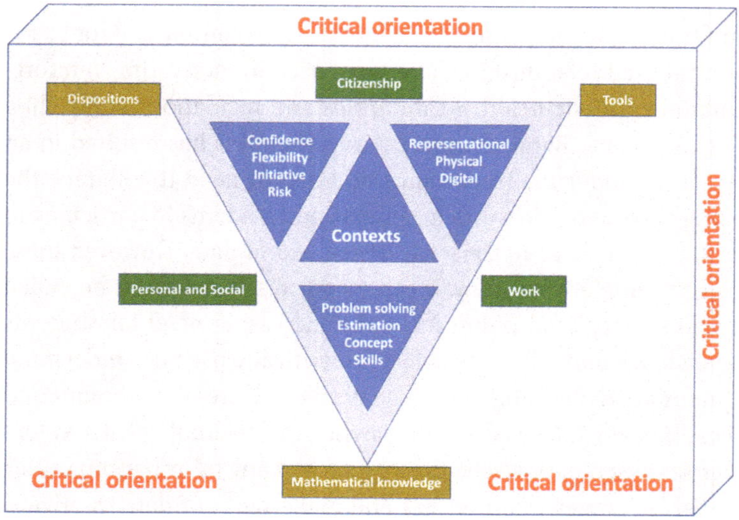

FIGURE 7.1 A framework for ML, recreated based on Geiger et al. (2014)

Niss & Jablonka, 2020); see also Chapter 1 in this book. One framework of ML by Goos and Geiger and colleagues (e.g., Geiger et al., 2014, 2015a, 2015b; Goos, 2007; Goos et al., 2013) was developed in a series of projects and publications as a practice- and research-based framework to be used as a tool for practice development, research, and teachers' professional development. These authors have constructed and used "a rich model of numeracy" (Geiger et al., 2015a, p. 612) based on the four intertwined elements or dimensions of *context, mathematical knowledge, tools,* and *dispositions*. These four dimensions are embedded within a *critical orientation* toward the application and use of mathematics in aspects of real-life related to the workplace (*work*), the *personal and social* sphere and *citizenship* (see Figure 7.1).

This ML framework depicts how mathematical knowledge in terms of applying mathematical concepts and (problem solving) skills is critically used to create meaning and solve problems from various contexts outside of school, which are of relevance for an individual's work-, personal- and social life, as well as the individual being an informed and responsible citizen. The framework also stresses the importance of the individual to be equipped with suitable *dispositions* toward mathematics in terms of having confidence in one's own mathematical competencies, and to willingly apply and engage in the flexible use of mathematics. In the framework, a *critical orientation* permeates all dimensions in which mathematical knowledge is used to understand, make, support or challenge arguments, judgements, and decisions.

In this chapter, we use the framework illustrated in Figure 7.1 as a reference framework to discuss SL and how the students' development of ML and SL can be understood, structured and supported using the models and modelling perspective on teaching and learning (e.g., Lesh & Doerr, 2003). Next, we set the stage to introduce SL by focusing on the discipline of statistics.

2 Statistics and Statistical Literacy

To be able to elaborate on the notion of SL in relation to ML, in this section we first discuss statistics as a discipline compared to the discipline of mathematics, as well as a framework for statistical thinking. Then we present some of conceptualizations of the notion of SL used in statistics education and relate these to our reference frame of ML.

2.1 *Statistics as a Discipline*

Within the field of statistics education, many authors stress the uniqueness of the discipline statistics, as well as the uniqueness of the challenges of teaching and learning statistics, compared to mathematics and mathematics teaching and learning (Cobb & Moore, 1997; del Mas, 2004; Rossman et al., 2006; Wild & Pfannkuch, 1999). Drawing on, and citing, the definition by the AMA – American Statistical Society – Wild et al. (2018) emphasise that "Statistics … is the science of learning from data, and of measuring, controlling and communicating uncertainty" (p. 6). A similar, but more terse definition is that statistics is "the science of gaining insight from data" (Rossman et al., 2006, p. 323). The focus on data, uncertainty and randomness are the three core features of statistics that are most often singled out as the defining differences between statistics and mathematics (Rossman et al., 2006). The general sentiment is that "[s]tatisticians are convinced that statistics, while a mathematical science, is not a subfield of mathematics" (Cobb & Moore, 1997, p. 814). All researchers do not agree with this view, however, and some do consider statistics as a branch or sub-discipline to mathematics. Capaldi (2019, p. 149), for example, defines mathematics as the "science of numbers: operations, interrelations, combinations, generalizations, and abstractions" and statistics as a "branch of mathematics: collection, analysis, interpretation, and presentation of masses of numerical data".

Although different researchers place different emphasis on what the nature of statistics is, most conceptualisations can to a large extent be linked and related to one or more aspects of the empirically-based four-dimensional framework of statistical thinking developed by Wild and Pfannkuch (1999). Indeed, in the teaching and learning of statistics there has been an ongoing shift during the last 30 years from a focus on skills, procedures, and computation towards SL, statistical reasoning, and statistical thinking (Ben-Zvi &

Garfield, 2004; Garfield & Ben-Zvi, 2008). In Ben-Zvi and Garfield (2004), the description of SL given entails (i) basic skills aimed at understanding statistical information such as organizing and representing data; and (ii) an understanding of the concepts, symbols and vocabulary used in the field of statistics. In comparison, statistical reasoning focuses on how statistical ideas are actually used to make sense of data and interpretations of statistical argument and results – with an emphasis on the actual reasoning being conducted. Lastly, statistical thinking focuses on the big ideas of statistics underlying statistical investigations, such as the various types of variation, the nature of samples and sampling, making inferences and critically evaluating statistical investigations and results (Ben-Zvi & Garfield, 2004). All three notions (SL, statistical reasoning, and statistical thinking) are interrelated in a fundamental way (cf. Wild & Pfannkuch, 1999; Pfannkuch & Wild, 2004), and the choice of perspective used in research depends on the focus of is the subject being studied.

2.1.1 Wild and Pfannkuch's Framework of Statistical Thinking

The empirically based four-dimensional framework of statistical thinking (ST) developed by Wild and Pfannkuch (1999), and further elaborated in Pfannkuch and Wild (2004), is composed of the four dimensions: (1) *the investigative cycle*; (2) *types of thinking*; (3) *the interrogative cycle*; and (4) *dispositions*. The first dimension is an adaptation of the so-called PPDAC model by MacKay and Oldsford (1994) that focuses on the actions and areas of attention when engaged in a statistical investigation. PPDAC stands for *Problem, Plan, Data, Analysis* and *Conclusion* and some of the characteristic actions and tasks involved in statistical investigations (see Figure 7.2). The investigative cycle is at the core of all statistical investigations since it anchors and centralises the whole statistical problem-solving process in the situation and phenomenon investigated – be it *personal, work related, social* or *societal* (*citizenship*); moreover, all five phases in the PPDAC model are fundamentally carried out in close interplay with the situation and phenomenon in question. Hence, the investigative cycle in Wild and Pfannkuch's statistical thinking framework can be seen as embodying the *context* dimension in the ML framework by Geiger et al. (2015a).

The second dimension, *types of thinking*, is divided into two categories: one gathering together general types of thinking applicable to all problem-solving situations, and another documenting the types of thinking that are specific for statistical problem solving. The different types of thinking are summarized in Table 7.1 and can be seen to describe the statistics-specific knowledge needed in applying and understanding, and doing and interpreting, statistics. Hence, in the context of the subject matter of statistics, the statistical thinking framework's dimension types of thinking correspond to the dimension of *mathematical knowledge* in the ML framework by Geiger et al. (2015a).

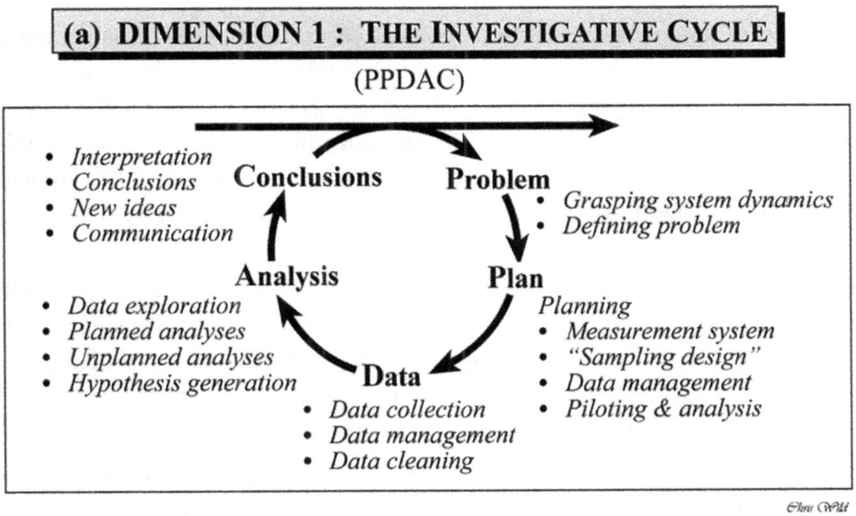

FIGURE 7.2 The investigative cycle (dimension 1) in Wild and Pfannkuch's *framework for statistical thinking* (Wild & Pfannkuch, 1999, p. 226)

TABLE 7.1 Types of thinking (dimension 2) in Wild and Pfannkuch's (1999) framework *for statistical thinking*: general types of thinking (G); and statistic-specific types of thinking (S).

Type of thinking		Characteristic of type of thinking
(G)	Strategic thinking	Planning, structuring, anticipating problems and challenges, awareness of (practical) constraints.
(G)	Modeling	Constructing models and as part of the investigative process.
(G)	Applying techniques	Using, adapting and applying existing statistical models and methods.
(S)	Recognition of need for data	The realization that anecdotes or personal experiences are not a sound foundation for answering statistical problems.
(S)	Transnumeration	Transforming and (dynamically) shaping and changing representations of data to better understand the system under investigation.
(S)	Consideration of variation	To acknowledge, account for and adequately handle uncertainty and variation.
(S)	Reasoning with statistical models	Knowing and using the distinctive set of models and reasoning skills required for statistical thinking.
(S)	Integrating the statistical and the contextual	Connecting and integrating the statistical models and reasoning to the particular context of a investigate study, an applied methodology and data.

The third dimension, *the interrogative cycle*, describes "a generic thinking process in constant use in statistical problem solving" (p. 231). The main objective of the interrogative cycle is to facilitate the distillation and encapsulation of the essence of the various inputs (ideas, data, information) given during the whole statistical problem-solving process. According to Wild and Pfannkuch, this process of distilling and encapsulating can be conceptualized as an iterative engagement involving the components *Generating*, *Seeking*, *Interpreting*, *Criticizing* and *Judging* (see Figure 7.3). In other words, the interrogative cycle is all about applying a critical stance towards all aspects of the statistical problem solving activity one engages in, and hence it parallels the *critical orientation* that permeates the ML framework by Geiger et al. (2015a).

The fourth dimension discusses a set of interrelated personal qualities and dispositions that are favorable with respect to statistical problem solving. As such, they correspond to the *disposition* dimension in the ML framework by Geiger et al. (2015a): *being curious and aware*; *being engaged*; *being imaginative*, *being skeptical*, *being logical*, and *having a propensity to seek deeper meaning while being open and exhibiting perseverance*:

- *Curiosity and awareness* here refer to being observant and cognizant of one's surroundings and willing to ask questions. This disposition often leads to new discoveries and, later, new insights.
- *Engagement* encapsulates having an interest in, or passion for, what one is doing, which facilitates discovering and noticing the small nuances and issues that may arise in the periphery of the subject one is involved in studying.
- *Imagination* is important to overcome stereotypical thinking, and involves seeing a problem from several different perspectives, as well as arriving at different, sometimes confusing, explanatory models and interpretations of data.

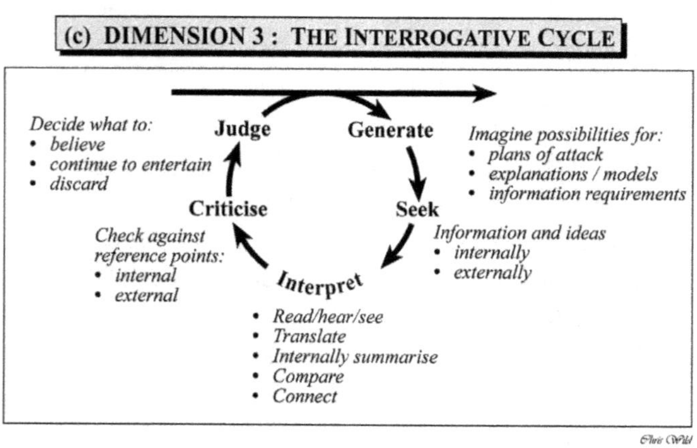

FIGURE 7.3 The interrogative cycle (dimension 3) in Wild and Pfannkuch's framework for statistical thinking (Wild & Pfannkuch, 1999, p. 226)

- *Skepticism* is about adapting a critical stance to methodological-, analytical- and interpretational flaws in the statistical problem-solving process. The essence of this disposition is manifested in the interrogative cycle.
- *Being logical* is a disposition that acts as a safeguard towards biases, and stresses that using logically valid and sound arguments and reasoning is the only way to arrive at valid conclusions and interpretations.
- *To a have a propensity to seek deeper meaning while being open and exhibiting perseverance* stresses an inert drive to look as far beyond the data as the problem situation and the data allow. *Perseverance* in this context refers to having patience with the complexity often involved in the situations and data analysed, and being *open* with a willingness to consider new ideas and possible explanations that potentially conflict with preconceived expectations, notions and relationships.

As indicated above, when interpreting Wild and Pfannkuch's (1999) framework of statistical thinking with respect to ML, their statistical thinking framework can in effect be taken as a model of SL where the *mathematical knowledge dimension* in the ML framework by Geiger et al. (2015a) has been replaced with a *statistical knowledge dimension.* Although this is a possibility, other more-or-less similar conceptualizations of statistical literacy have also been discussed in the literature, which we will now briefly discuss next.

2.2 *Statistical Literacy*

In the same way that statistics education research does not consider statistics a sub-discipline of mathematics, SL, statistical reasoning, and statistical thinking are also stressed as not identical to ML, mathematical reasoning, and mathematical thinking (Ben-Zvi & Garfield, 2004). However, as in the cases of the varied meanings and uses of the notion of ML in the mathematics educational literature (Geiger et al., 2015a; Jablonka, 2003; Niss & Jablonka, 2020), the same variation can be observed in the notion of SL (Sharma, 2017). For example, Ben-Zvi and Garfield (2004) note that SL is often used interchangeably with quantitative literacy (c.f. Steen, 1999), and Gould (2017) argues that data literacy (c.f. ODI, 2015) can in essence be understood as SL. A reason for why there are many competing and complementing definitions and frameworks of SL might be seen as summed up in the following quote by Watson and Callingham (2020): "Statistical literacy is the meeting point of the practice of statistics and the everyday world" (p. 19). The specificities of the 'meeting points' that are considered can naturally be vastly different in nature and with respect to what aspects are coming to the fore, as well as important in different contexts where statistical thinking and statistical problem solving are applied.

Hence, it is not surprising that the different characteristics of SL expressed in the research literature are coloured by the context, situation, and problems where the statistics is used to create meaning and understanding.

A recent literature review by Sharma (2017) discusses the different definitions and frameworks of SL. Here, we will partly draw on Sharma's work to highlight statistical literacy as discussed by Gal (2000, 2002) and Watson and Callingham (2003), respectively, by situating these with respect to the ML framework by Geiger et al. (2015a). Note that *probability literacy* has also been discussed in the literature (Gal, 2005), but due to space limitation, we do not address this notion in the present chapter.

2.2.1 Two Frameworks of Statistical Literacy

In her presidential address, *Enhancing statistical literacy: Enriching our society*, at the 1992 annual meeting of the American Statistical Association, Wallman presented the following definition:

> "Statistical literacy" is the ability to understand and critically evaluate statistical results that permeate our daily lives – coupled with the ability to appreciate the contributions that statistical thinking can make in public and private professional and personal decisions. (Wallman, 1993, p. 1)

This sentiment expressed from within the statistics discipline itself has been built on and adapted in the more recent field of statistics education in various ways. Indeed, the framework of Wild and Pfannkuch (1999) can be seen as an extended elaboration of the key aspects in Wallman's definition, as can the SL framework proposed by Gal (2000, 2002). Gal argues that the two primary components of SL that educated adults need to have in order to prosper and be productive in all facets of their lives are:

> (a) people's ability to interpret and critically evaluate statistical information, data-related arguments, or stochastic phenomena, which they may encounter in diverse contexts, and then relevant (b) their ability to discuss or communicate their reactions to such statistical information, such as their understanding of the meaning of the information, their opinions about the implications of this information, or their concerns regarding the acceptability of given conclusions. (Gal, 2002, pp. 2–3)

Gal elaborates these two aspects by putting forward a set of knowledge- and dispositional elements needed to be developed to become statistically literate: five (*statistical*) *knowledge elements* (SKEs) and two (*statistical*) *dispositional elements* (SDEs). These seven elements are listed in Table 7.2.

TABLE 7.2 The seven knowledge and disposition elements in Gal's (2002) framework of SL

Statistical literacy elements	Some characteristics of the element
(SKE1) Literacy skills	General text-processing skills (reading/writing) of different types of texts in different contexts and with different intentions.
(SKE2) Statistical knowledge	A solid understanding and knowledge about basic concepts, representations and methods in statistics and probability.
(SKE3) Mathematical knowledge	Basic mathematical concepts and methods, in particular number sense, logical derivations and reasoning, notation and representations.
(SKE4) Context knowledge	To understand that data/numbers in statistics are 'Numbers in context' and hence knowledge about the context is needed to make sense of data, analyses and results.
(SKE5) Critical questions	Knowing how to be critical and ask questions about data, methods, results and how these are presented, as well as motivation and possible agendas for a study and chosen perspectives/angles.
(SDE1) Beliefs and attitudes	Positive beliefs and attitudes towards the possibilities and power in using statistics as well as the ability and willingness to engage in/with statistics and take a critical stance.
(SDE2) Critical stance	Being cautious towards one's first impressions and being aware of one's own (and others') biases.

In terms of ML, the characterization of SL presented by Gal (2005) explicitly and clearly stresses the importance of [*mathematical*] *knowledge*, which is evident from (SKE2) and (SKE3). It can also be seen that *tools* are represented in (SKE3), emphasizing notion and representations. The importance and dependence of *context* is stressed foremost in (SKE4) but also in (SKE1), whereas (SDE1) entails *productive dispositions. Critical orientation* can be seen as expressed in four of the seven elements; explicitly in (SDE1) and (SKE5) as having a critical stance, and also in (SDE2) in terms of vigilance towards one's own biases, and as a component of (general) literacy skills (SKE1). Hence, Gal's SL can by and large be understood as ML, where the *mathematical knowledge dimension* includes a clear focus on the content matter related to statistical thinking (cf. Wild & Pfannkuch, 1999). Note that Gal's definition of SL to various degrees connects more or less explicitly to other types of literacy: (SKE1) to document literacy (Kirsch et al., 1998) and critical literacy (Luke & Freebody,

1997); (SKE2) to probability literacy (Gal, 2005); and (SKE3) to the knowledge dimension in ML (Geiger et al., 2015a).

In trying to operationalise these two primary fundamental components of SL, as discussed by Gal (2000, 2002), to support students in becoming statistically literate, Watson and Callingham (2003), drawing on Watson (1997) and Gal (2000), suggested a three-tiered framework for SL including "(i) the understanding of basic statistical terminology, (ii) the understanding of terminology when it appears in social contexts, and (iii) the ability to question claims that are made in context without proper statistical justification" (p. 7). They further presented a six-level model of SL in relation to the expectations for what individual students may be competent to express and apply when they leave school to participate in society (see Table 7.3).

TABLE 7.3 The SL construct by Watson and Callingham (2003) as summarized by Sharma (2017, p. 123)

		Some characteristics of the element
6.	Critical Mathematical	Critical, questioning engagement with context, using proportional reasoning particularly in media or chance contexts, showing appreciation of the need for uncertainty in making predictions, and interpreting subtle aspects of language.
5.	Critical	Critical, questioning engagement in familiar and unfamiliar contexts that do not involve proportional reasoning, but which do involve appropriate use of terminology, qualitative interpretation of chance, and appreciation of variation.
4.	Consistent Non-critical	Appropriate but non-critical engagement with context, multiple aspects of terminology usage, appreciation of variation in chance settings only, and statistical skills associated with the mean, simple probabilities, and graph characteristics.
3.	Inconsistent	Often in supportive formats, expect selective engagement with context, appropriate recognition of conclusions but without justification, and qualitative rather than quantitative use of statistical ideas.
2.	Informal	Only colloquial or informal engagement with context often reflecting intuitive non-statistical beliefs, single elements of complex terminology and settings, and basic one-step straightforward table, graph, and chance calculations.
1.	Idiosyncratic	Engagement with context and use of terminology, basic mathematical skills in an idiosyncratic way.

Watson (2011) argues that these six levels are crucial in the progression toward SL, regardless of the age of the students, and have been used to design and test SL (Watson & Callingham, 2003; Berndt et al., 2021). In the next section, we will consider another approach to support and facilitate students in becoming statistically-, as well as mathematically-, literate.

3 The Models and Modelling Perspective and Its Relation to Mathematical Literacy and Statistical Literacy

In this section, we will introduce the models and modelling perspective on teaching and learning (Lesh & Doerr, 2003), and relate this to, and briefly discuss how, it can facilitate the design of learning opportunities supporting students in developing ML and SL.

3.1 *The Models and Modelling Perspective*

The models and modelling perspective on teaching and learning (Lesh & Doerr, 2003) is a theoretical framework in which a *model* is defined as a general system consisting of elements, relationships, rules, and operations that can be used to make sense of, predict, describe, or explain some other system. *Mathematical models* specifically focus on the structural characteristics of the system in question, and the core of engaging in *mathematical modelling* is the use of mathematical models as sense-making tools. The foundation of the models and modelling perspective is based on the ideas of Vygotsky, Piaget and Dienes as well as the American pragmatist tradition represented by Mead, Peirce and Dewey (Lesh & Doerr, 2003). As one of many perspectives on mathematical modelling in the literature, the models and modelling perspective is an example of a *contextual perspective* (Kaiser & Sriraman, 2006) and has both subject-matter related goals (e.g., "structuring of learning processes, introduction of new mathematical concepts and methods including their illustration" (p. 302)) as well as psychological goals (e.g., "fostering and enhancement of the motivation and attitude of learners towards mathematics and mathematics teaching" (pp. 302–303)).

3.1.1 Model Development Sequences

To support and facilitate students' development of models towards a given learning goal, the models and modelling perspective provides a guiding framework for the purposeful design and structuring of learning in terms of three different types of structurally-related activities organized in *model development sequences*. These activities are as follows: (i) *model eliciting activities* aimed at eliciting the ideas the students bring to the activity; (ii) *model exploration activities* which focus on exploring and developing the underlying mathematical structure of the models elicited by students; and (iii) *model application activities* in which students apply their developed models in similar or new contexts. Common to the three types of activities are that they fundamentally involve students to iteratively engage in exploring, expressing, testing, revising, and developing their models to make sense of different situations and contexts (Lesh et al., 2003; Lesh & Doerr, 2003).

3.1.2 Model Eliciting Activities

Purposefully designed model eliciting activities, which can be used as standalone activities to introduce a particular learning objective or as the first activity in a model development sequence, have been used in a range of disciplines and contexts to support and investigate students' learning (Diefes-Dux et al., 2006; Iversen & Larsen, 2006; Yildirim et al., 2010; Yoon et al., 2010). The central idea with a model eliciting activity is to present students to a context and a problem where they need to develop a model that can be used to describe, explain or predict the behaviour of some situation or phenomenon (Lesh et al., 2000). Lesh and colleagues (2000) have developed the following six principles to guide the design of model eliciting activities: (a) *the reality principle* – the context and problem posed in the model eliciting activity should be meaningful to the students and connect to their previous experiences; (b) *the model construction principle* – that the students are faced with an inherent need to develop a model when engaging in the activity; (c) *the model documentation principle* – that the situation in which the students are working with the model eliciting activity requires the students to express the models they construct and use in terms of some form of external representation; (d) *the model generalization and sharable principle* – that the models elicited in the model eliciting activity are neither wholly private nor too narrow in scope, but rather sharable with peers, generalizable and applicable to similar situations; (e) *the simplicity principle* – that the situation in, as well as the formulation of, the model eliciting activity is as simple as possible in order to facilitate that a particular model (learning goal) is targeted; and (f) *the self-evaluation principle* – that the model eliciting activity permits the students to by themselves assess and evaluate their models and work (Lesh et al., 2000; Lesh & Doerr, 2003). Much of the research done from a models and modelling perspective focus on model eliciting activities, whereas research targeting model exploration activities or model application activities are more spares. Hence, design principles are not well established or elaborated for these two latter types of activities.

3.2 *Mathematical Literacy and Statistical Literacy from a Models and Modelling Perspective*

We argue that the tenet of the models and modelling perspective, that mathematical models are sense-making tools used to understand, predict, describe, or explain phenomena in difference contexts, fundamentally resonates with the dimensions of ML and elements of SL. In the models and modelling perspective, the *critical orientation* stressed in ML and SL underlines and permeates how models are developed and applied in the models and modelling perspective by students actively and iteratively engaging in exploring, expressing, testing, revising, and developing their models to make sense of different *situations and contexts*. The *reality design principle* for model eliciting activities naturally supports the inclusion of problems from the *private-, social-*, (future) *work related-* and *societal domains*, as do *model application activities*. The model eliciting activity *model generalization and sharable design principle* in addition helps to focus on more general models that are applicable in different contexts. Further, as characterized as a contextual perspective, the models and modelling perspective has both *mathematical knowledge* related- and *dispositional* goals (cf. Kaiser & Sriraman, 2006). The *mathematical knowledge* in the models and modelling perspective is systematically and dialectally being developed and applied through model development sequences, and the model eliciting activity *self-evaluation design principle* strives to support students in developing productive and sound *dispositions* toward using and developing mathematics to understand phenomena and solve problems. The use of *tools* is an integral part of all the activities in model development sequences, but in particular they come to the fore in model exploration activities, where various forms of media, concreate materials and tools are used to explore and develop different multimodal representations of the mathematical concepts or models in use.

For us, the models and modelling perspective provides model eliciting activities and model development sequences as didactical tools for structuring students' learning opportunities and supporting them in iteratively developing their sense-making mathematical models with a special emphasis on exposing the students to different relevant phenomena and contexts in which to apply their models. We broadly see models as 'the glue' that connects the dimensions in the ML framework by Geiger et al. (2015a), since it is through the lens of the constructed and used models that the coordinated conglomerate of *context, mathematical knowledge, tool use*, and *disposition* results in what the individual *critically* understands and potentially acts upon. We illustrate this model-centred perspective of ML in Figure 7.4.

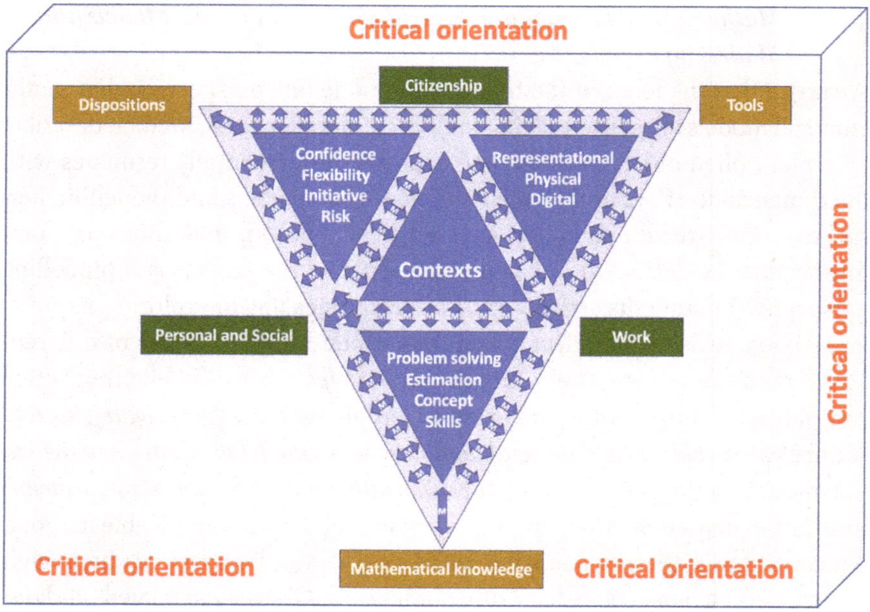

FIGURE 7.4 Arrows added to Figure 7.1 to illustrate how models (cf. Lesh & Doerr, 2003) fundamentally permeate and hold together the dimension and components of ML (building on the framework by Geiger et al. (2014))

We now use the conceptualization represented in Figure 7.4 to illustrate students' opportunities to develop their ML and SL by presenting an example of an activity from our research on the teaching and learning of statistics at the lower secondary level, were we used the models and modelling perspective as a framework for both design and analysis.

4 How Long Is a Minute? – An Activity Supporting Students in Developing Mathematical Literacy in the Contexts of Statistics

This example is centred around a modelling activity within the contexts of time, beats per minutes and pulse,[2] which comes from a model development sequence aimed at covering the statistical content matter in the Swedish mathematics curriculum for Grade 7 (Skolverket, 2012). The activity is a *model application activity* in which the students apply the central measures of mean and median to explore and experience their sense of time in an applied situation by engaging in a small statistical investigation. From a ML and SL perspective the *mathematical-* and *statistical knowledge* in focus in the activity is the content of mean and median. In addition, the activity involves formulating simple hypotheses and evaluating these hypotheses by collecting and analysing data, which are connected to three of the phases of the PPDAC model of statistical investigation and components of statistical literacy (cf. (SKE2) in Gal's (2002) framework). Hence, the learning goal for the activity was for the students to both conduct and participate in a small empirical statistical investigation, and to apply the central measures of mean and median in, what was for the students, a new context.

The class in which the activity was implemented had a music profile, meaning that the class had extra time in their schedule for various music lessons, practice, activities, and performances. According to the teacher however, the class was an average class with respect to their knowledge in, and attitudes towards, mathematics. The mathematics teacher of the class was also the students' music teacher.

When introducing the activity and waiting for the PowerPoint presentation to start, the teacher told the students that they were going to work in pairs and collect some data. He asked them if they remembered when they talked about BPM (Beats Per Minutes) in music class, and the importance of both "feeling the beat" and keeping the beat constant when playing together. When the presentation started, he read "How long is a minute?" on the opening slide, and some students immediately responded by saying "60 seconds". The teacher acknowledged this and continued to start a metronome-app on his smartphone set at 60 BPM and evenly and rhythmically in synchronization with the metronome said "this-is-how-long-a-second-is". In this way, the teacher aimed to elicit the students' dispositions of curiosity, awareness and engagement (cf. components of the fourth dimensions of Wild and Pfannkuch's statistical thinking framework and ML) by motivating and capturing the students' interest by connecting the coming task to multiple relevant contexts and familiar situations for the students.

The teacher continued to discuss the notions of beat, pulse and BPM with the students. He did this using various contexts: an electrocardiogram (EKG); how to measure your own pulse; an exercise heart rate monitor; people jogging; an analogue- and digital metronome; and, a piece of sheet music which included a set tempo marking (♪ = 72)[3]; see Figure 7.5. In terms of ML and SL, this focus on *dispositions, beliefs and attitudes* in the introduction of the activity served to prepare and prime the students' expectations and willingness to engage mathematically and statistically with the situation to come.

The teacher then transitioned to explain how the students could use the stopwatch function on their smartphones as a *tool* to take measurements when their peer was trying to estimate how long a minute is in three different circumstances. Tool use is a dimension in the framework for ML by Geiger et al.

FIGURE 7.5 The slide the teacher used in discussing the notions of beat, pulse, and BPM

(2014) as well as a component stressed in Wild and Pfannkuch's (1999) framework for statistical thinking. More precisely, the activity presented and shown on a slide to the students was as follows:

We are going to compare how long you think a minute is if:

(1) you are relaxed and at rest in a calm and quiet environment;
(2) you are in an environment with "disruptive" music[4] [i.e., quite loud music with modern harmonization, non-trivial rhythms and tempi changes];
(3) you have just completed 20 rapid jumping-jacks.

In the planned flow of the activity, the teacher here was supposed to engage the students in expressing their hypotheses regarding these three tasks. Formulating hypotheses to critically investigate with collected data is central to the statistical problem-solving process and coming to the fore in all SL frameworks previously discussed in the chapter. Having articulated hypotheses facilitates taking a critical stance towards the whole process and asking critical questions, which is an important dimension of ML as well as a part of SL (cf. Gal (2002): (SKE5) – Critical questions; (SDE2) – Critical stance). However, some clarifying questions from the students seemingly threw the teacher off script, so he forgot this part of the planning, and instead the class started engaging in measuring how long they thought a minute was in the three scenarios.

When the students had collected all their data, the teacher sent a sheet of paper around the class to collect all data in a single document, and at the same time initiated and led a whole-class discussion about the students' hypotheses, which he had previously forgotten. He started by again stressing the importance of stable beats when making music together, before having a dialog with the students regarding (a) in which of the three scenarios it was the hardest to estimate how long a minute is and why they thought so; (b) in which of the three scenarios their estimates deviated the most from 60; and (c) whether they were surprised by their results. All three lines of questioning to engage in different types of thinking (cf. dimension 2 in Wild and Pfannkuch's (1999) framework for statistical thinking) and encouraged the students to express and become explicitly aware of their first spontaneous ideas, impressions, and potential biases regarding the three scenarios, making these object to evaluation and critique. Indeed, the students offered many experiences and hypotheses illustrating how they consider the variation in the three scenarios: the relaxed and quiet scenario was difficult since it was challenging to not lose one's concentration and keep focus on the task; the noisy music with mixed rhythms disturbed the even pace of counting from 1 and 60, and the increased pulse and being somewhat out of breath made one count faster. Some students who were very active in sports presented the idea that they were not affected so much in their performance in estimating a minute, since the physical exertion did not affect their pulse very much. Related to this line of reasoning, the teacher connected this to other sports where being aware of one's pulse is important and explicitly talked about, such as biathlon. Related to the students' comments on how the noisy music with mixed rhythms disturbed the even pace of counting, the teacher broadened the discussion to talk about distractions in general concerning different types and contexts: loud noises due to traffic; distracting talk [in the school setting]; ongoing construction work; bad lighting; or getting sun in one's eyes. This discussion, which in various ways stressed the *contexts* of the three scenarios in addition highlighted students' model thinking and their integrating of the statistical element with context (cf. dimension 2 in Wild and Pfannkuch's (1999) framework for statistical thinking), ended the first part of the activity and the lesson.

The activity was resumed in the next lesson, with the teacher asking the students what scenario affected them the most and the least with respect to estimating how long a minute is. The students again expressed different opinions, and the teacher asked them to justify their thinking in order to encourage the students to be logical in the sense of Wild and Pfannkuch's (1999) fourth dimension, of interrelated personal qualities and dispositions, in their framework for statistical thinking. One student said that the exertion made her feel her pulse more clearly and hence she could use that as a reference beat when counting the seconds. Someone then commented that the reference beat might result in counting too fast, and many students concurred that counting too fast was what they experienced. The teacher connected the increase in pulse to being in some sort of distress and asked if this affect could also be the result of other stimuli, like noise and disturbing music. This suggestion of the teacher offered the students to consider the model of the phenomena of an increasing pulse to be a more general model applicable also in other situations and contexts. To this suggestion students both agreed and disagreed. One group of students argued that the music is like 'white noise' which helped them to really focus on the counting of seconds, whereas total silence can distract you even by the smallest of sounds. The consensus expressed by the class was that music can help you focus, but then it is important that it is the right type of music. This whole class discussion again focused on the three different *contexts* of the scenarios and to think critically about these and both their initial thinking and developing reasoning.

The mathematical- and statistical knowledge and content in the activity then came to the fore with the help of the calculations of two students. The teacher presented the compiled results from the data from all pairs on a smartboard (see Table 7.4), and focusing on the columns 'Rest' and 'Noise' asked, by pointing at the table: "Can anyone explain this result, that it's [the mean] going down here [from 64 to 61], and that it's [the median] going up here [from 61 to 63]?". Here, the teacher asked the students to make sense of the applied mathematical and statistical constructs, procedures and representations. Both from a ML- and SL perspective the interplay between the context of the situation and the mathematics and statistics used is key to create a model to make sense of the problem at hand. The question posed by the teacher focuses the students' attention to think about the data as numbers in context (cf. Gal's (2002): (SKE4) – Context knowledge), which in addition facilitate taking a critical stance in interpreting their empirical results.

First, some of the students said "No, that's unexplainable"; others again returned to pointing out the differences in the three scenarios, so the teacher had to repeat the question and explicitly ask why the mean and the median points were different regarding in what scenario ('rest' or 'noise') the students

TABLE 7.4 The summary statistics for the class data as presented by the teachers on the smartboard

Measure	Rest (sec)	Noise (sec)	Exertion (sec)
Mean	64	61	62
Median	61	63	60

best estimated how long a minute was. One student then offered the explanation that people are different both when it comes to preferences and their conditions, and that these can be factors explaining the result. The teacher acknowledged this source of variation and used it to push the students to think about the constructs of mean and median, and how these are calculated. With the support of the students, he briefly recapitulated how means and medians are calculated, and then asked: "When is it good to use mean and when is it good to use median when we make investigations?" This triggered a discussion about the distribution of the underlying data and how much effect outliers potentially have on the mean and the median. Here, the teacher used an on-the-spot made up example about money distribution. By turning to the representation of data in terms of the central measures of mean and median in the table, the mathematical- and statistical knowledge relevant in the activity became the focus of the students' attention. As a result, (a) the crucial role of data in trying to understand the phenomena was highlighted; (b) the important statistical idea of variability surfaced and was connected to variation in the contexts (differences between individuals) in which the data were collected; (c) the mathematical knowledge about how to calculate means and medians were explicitly recapitulated by, and to, the students; and (d) critical questions regarding how the central measures of means and medians might depend on the distribution of the collected data were addressed. In other words, the activity and the teacher's managing of the classroom discussions provided the students with opportunities to engage, express and progressively elaborate multiple levels of thinking and reasoning as specified according to Watson and Callingham's (2003) model of SL.

After having pointed out that, according to the calculated means and medians, the students in the class were rather good at estimating how long a minute is, the teacher asked the students if they could draw any certain conclusion based on their collected data and analysis. After some discussion, they reached the conclusion that they could not, at least not just based on the summary statistics presented in the table. However, they argued that looking back at the data, and taking outliers into account might provide a clearer answer. On class inspection of the data, an outlier of 84 seconds in the exertion scenario was found, and the implication of this on the mean- and median values was explored. Just before the activity ended with the class agreeing that in this particular case the mean provided a more accurate picture of the data, a student also pointed out that the numbers in the table were rounded numbers and that if the full accuracy (using hundredths of a second) was used, the difference between the scenarios might be much smaller. Hence, the activity ended with the teacher further encouraging the students to take a critical look at the small investigation they had conducted. In the discussion, the consequences of variability in the data in relation to the used mathematics and representations to make sense of the results in the three contexts of the scenarios was stressed. In addition, a critical question regarding the applied method of data collection was raised. Statistical investigation, data collection, contextual dependence of data, data variability, and a critical disposition are all specific components of SL, and more generally of ML, which the students engaged in and developed throughout the activity.

At the very end of the model development sequence on statistics, the teacher summarised all the work the students had done and connected it back to the activity just discussed by showing a short (one-minute-long!) YouTube video[5] giving examples of what transpires in one minute. This video puts different contexts and the frequency of phenomena from various areas in life, the workplace and society into the perspective of a 60 second timespan. The video hence provided the students with a possibility to see how the mathematical- and statistical knowledge used and discussed in the activity could be relevant, connected and applicable to other situations they might encounter.

5 Summary

In this chapter we have elaborated on the learning of statistics from an ML point of view. Although mathematics and statistics can be seen as two different, yet related, disciplines (Cobb & Moore, 1997), we have provided a perspective on bringing the teaching and learning of mathematics and statistics closer together, as summarized in Figure 7.4. We have used an example from our research to illustrate how the models and modelling perspective and model development sequences can facilitate the design and implementation of learning opportunities to support students in developing SL and ML. This example illustrates how the teacher strived to develop a meeting point of the practice of statistics and the everyday world (Watson & Callingham, 2020) by using the students' own interest as a departure for the activity. The activity did not just end with a discussion about means, medians and outliers, it also gave an authentic and realistic context for the students to attach to these discussed statistical concepts. Throughout the activity the students were provided with opportunities to engage in different types of statistical thinking and reasoning (Watson & Callingham, 2003; Wild & Pfannkuch, 1999), such as to interpret and critically evaluate statistical information, make hypotheses, and discuss and criticize results, all of which are central elements of SL (Gal, 2000, 2002) and ML (Geiger et al., 2015).

Based on our discussion in the chapter, we argue that statistics, as a form of mathematical content in school mathematics curricula, is the one content where all dimensions of ML most naturally and clearly come to the fore. Hence, we think that connections between ML and SL should be more systematically investigated, and in such an endeavour, this chapter shows that the models and modelling perspective and model development sequences are promising tools for future research.

Notes

1 Niss and Jablonka (2020, p. 549) list the following *numeracy, quantitative literacy, critical literacy, mathemacy, matheracy* and *statistical literacy.*
2 In Swedish the corresponding words "puls" ['pulse' in English] and "takt" ['beat' in English] are often used interchangeable in most (everyday) discourses – and especially in music.
3 This notation tells that the duration of an eighth note should corresponded to one 72nd of a second (72 BPM).
4 The music played was Frank Zappa's *What's new in Ballimore* from the album *Frank Zappa meets the brothers of prevention.*
5 https://youtu.be/zhWDdy_5v2w

References

Aguilar, M. S., & Castaneda, A. (2021). What mathematical competencies does a citizen need to interpret Mexico's official information about the COVID-19 pandemic? *Educational Studies in Mathematics, 108*(3), 227–248. https://doi.org/10.1007/s10649-021-10082-9

Ben-Zvi, D., & Garfield, J. (2004). Statistical literacy, reasoning, and thinking: Goals, definitions, and challenges. In D. Ben-Zvi & J. Garfield (Eds.) *The challenges of developing statistical literacy, reasoning and thinking* (pp. 3–15). Kluwer Academic Publishers.

Berndt, M., Schmidt, F. M, Sailer, M., Fischer, F., Fischer, M. R., & Zottmann, J. M. (2021). Investigating statistical literacy and scientific reasoning & argumentation in media-, social sciences-, and economic students. *Learning and Individual Differences, 86,* Article id: 101963. https://doi.org/10.1016/j.lindif.2020.101963

Capaldi, M. B. (2019). Mathematics versus statistics. *Journal of Humanistic Mathematics, 9*(2), 149–156. https://doi.org/10.5642/jhummath.201902.10

Cobb, G. W., & Moore, D. S. (1997). Mathematics, statistics, and teaching. *The American Mathematical Monthly, 104*(9), 801–823. https://doi.org/10.2307/2975286

del Mas, R. C. (2004). A comparison of mathematical and statistical reasoning. In D. Ben-Zvi & J. Garfield (Eds.), *The challenge of developing statistical literacy, reasoning, and thinking* (pp. 79– 96). Kluwer Academic Publishers.

Diefes-Dux, H., Bowman, K., Zawojewski, J., & Hjalmarson, M. (2006). Quantifying aluminum crystal size part 1: The model eliciting activity. *Journal of STEM Education: Innovations and Research*, *7*(1&2), 51–63.

Franklin, C., Kader, G., Mewborn, D., Moreno, J., Peck, R., Perry, M., & Scheaffer, R. (2007). *Guidelines for Assessment and Instruction in Statistics Education (GAISE) Report: A Pre-K–12 Curriculum framework*. American Statistical Association.

Gal, I. (2000). Statistical literacy: Conceptual and instructional issues. In D. Coben, J. O'Donoghue, & G. E. Fitzsimons (Eds.), *Perspectives on adults learning mathematics: Research and practice* (pp. 135–150). Kluwer Academic Publishers.

Gal, I. (2002). Adults' statistical literacy: Meanings, components, responsibilities. *International Statistical Review*, *70*(1), 1–25. https://doi.org/10.2307/1403713

Gal, I. (2005). Towards "probability literacy" for all citizens: Building blocks and instructional dilemmas. In G. A. Jones (Ed). *Exploring probability in school: Challenges for teaching and learning* (pp. 39–65). Springer. https://doi.org/10.1007/0-387-24530-8_3

Gardner, M (1970). The paradox of the nontransitive dice. *Scientific American 223*(6), 110–111.

Gardner, M. (1983). *Wheels, life, and other mathematical amusements*. Freeman.

Garfield, J. B., & Ben-Zvi, D. (2008). *Developing students' statistical reasoning: Connecting research and teaching practice*. Springer.

Geiger, V., Goos, M., & Dole, S. (2014). Students' perspectives on their numeracy development across the learning areas. In Y. Li & G. Lappan (Eds.), *Mathematics curriculum in school education* (pp. 473–492). Springer.

Geiger, V., Goos, M., & Forgasz, H. (2015a). A rich interpretation of numeracy for the 21st century: A survey of the state of the field. *ZDM Mathematics Education, 47*(4), 531–548. https://doi.org/10.1007/s11858-015-0708-1

Geiger, V., Goos, M., & Forgasz, H. (2015b). A critical orientation to numeracy across the curriculum. *ZDM Mathematics Education, 47*(4), 611–624. https://doi.org/10.1007/s11858-014-0648-1

Goos, M. (2007, August). *Developing numeracy in the learning areas (middle years)*. Keynote address delivered at the South Australian Literacy and NumeracyExpo, Adelaide, Australia.

Goos, M., Geiger, V., & Dole, S. (2013). Designing rich numeracy tasks. In C. Margolinas (Ed.), *Task design in mathematics education. Proceedings of ICMI Study 22* (vol. 1). Oxford.

Gould, R. (2017). Data literacy is statistical literacy. *Statistics Education Research Journal, 16*(1), 22–25.

Heyd-Metzuyanim, E., Sharon, A. J., &, Baram-Tsabari, A. (2021). Mathematical media literacy in the COVID-19 pandemic and its relation to school mathematics education. *Educational Studies in Mathematics, 108*(3), 201–225. https://doi.org/10.1007/s10649-021-10075-8

Iversen, S., & Larson, C. (2006). Simple thinking using complex math vs. complex thinking using simple math – A study using model eliciting activities to compare students' abilities in standardized tests to their modelling abilities. *ZDM Mathematics Education, 38*(3), 281–292. https://doi.org/10.1007/BF02652811

Jablonka, E. (2003). Mathematical literacy. In A. Bishop, M. A. Clements, C. Keitel, J. Kilpatrick, & F. S. K. Leung (Eds.), *Second international handbook of mathematics education* (pp. 75–102). Kluwer Academic Publishers.

Jablonka, E., & Bergsten, C. (2021). Numbers don't speak for themselves: Strategies of using numbers in public policy discourse. *Educational Studies in Mathematics, 108*(3), 579–596. https://doi.org/10.1007/s10649-021-10059-8

Kaiser, G., & Sriraman, B. (2006). A global survey of international perspectives on modelling in mathematics education. *ZDM Mathematics Education, 38*(3), 302–310. https://doi.org/10.1007/BF02652813

Kirsch, I. S., Jungeblut, A., & Mosenthal, P. B. (1998). The measurement of adult Literacy. In S. T. Murray, I. S. Kirsch, & L. B. Jenkins (Eds.). *Adult literacy in OECD countries: Technical report on the first international adult literacy survey* (pp. 105–134). National Center for Education Statistics, U.S. Department of Education.

Lesh, R., Hoover, M., Hole. B., Kelly, E., & Post, T. (2000). Principles for developing thought-revealing activities for students and teachers. In A. Kelly & R. Lesh (Eds.), *Handbook of research design in mathematics and science education* (pp. 591–646). Lawrence Erlbaum Associates.

Lesh, R., Cramer, K, Doerr, H. M., Post, T., & Zawojewski, J. S. (2003). Model development sequence. In R. Lesh & H. M. Doerr (Eds.), *Beyond constructivism: Models and modeling perspectives on mathematics problem solving, learning, and teaching* (pp. 35–58). Lawrence Erlbaum Associates.

Lesh, R., & Doerr, H. M. (Eds.). (2003). *Beyond constructivism: Models and modeling perspectives on mathematics problem solving, learning, and teaching.* Lawrence Erlbaum Associates.

Luke, A., & Freebody, P. (1997). Shaping the social practices of reading. In S. Musprati, A. Luke, & P. Freebody (Eds.), *Constructing critical literacies: Teaching and learning textual practice* (pp. 185–225). Allen & Unwin.

MacKay, R. J., & Oldford, W. (1994). *Stat 231 course notes full 1994.* University of Waterloo.

Manyika, J., Chui, M., Brown, B., Bughin, J., Dobbs, R., Roxburgh, C., & Hung Byers, A. (2011). *Big data: The next frontier for innovation, competition and productivity.* McKinsey Global Institute.

Niss, M., & Jablonka, E. (2020). Mathematical literacy. In S. Lerman (Ed.), *Encyclopedia of mathematics education* (pp. 548–553). Springer.

Oceans of Data Institute (ODI). (2015). *Building global interest in data literacy: A dialogue.* Educational Development Center.

Organisation for Economic Co-operation and Development (OECD). (2013). *PISA 2012 assessment and analytical framework. Mathematics, reading, science, problem solving and financial literacy.* OECD.

Pfannkuch, M., & Wild, C. (2004). Towards an understanding of statistical thinking. In D. Ben-Zvi & J. Garfield (Eds.), *The challenges of developing statistical literacy, reasoning and thinking* (pp. 3–15). Kluwer Academic Press.

Rossman, A. J., Chance, B. L., & Medina, E. (2006). Some important comparisons between statistics and mathematics, and why teachers should care. In G. F. Burrill (Ed.), *Thinking and reasoning about data and chance: Sixty-eighth NCTM yearbook* (pp. 323–334). National Council of Teachers of Mathematics.

Rump, C. M. (2001). Strategies for rolling the Efron dice. *Mathematics Magazine, 74*(3), 212–216. https://doi.org/10.2307/2690722

Savage, R. P. (1994). The paradox on nontransitive dice. *The American Mathematical Monthly, 101*(5), 429–436. https://doi.org/10.2307/2974903

Sharma, S. (2017). Definitions and models of statistical literacy: A literature review. *Open Review of Educational Research, 4*(1), 118–133. https://doi.org/10.1080/23265507.2017.1354313

Skolverket. (2012). *The subject syllabus for mathematics.* https://www.skolverket.se/download/18.4fc05a3f164131a74181063/1535372298267/Mathematics-swedish-school.pdf

Steen, L. A. (1999). Numeracy: The new literacy for a data-drenched society. *Educational Leadership, 57*(2), pp. 8–13.

United Nations Environment Programme (UNEP). (2020a). *How COVID-19 is changing the world: A statistical perspective.* UN Document Repository. https://unstats.un.org/unsd/ccsa/documents/covid19-report-ccsa.pdf

United Nations Environment Programme (UNEP). (2020b). *How COVID-19 is changing the world: A statistical perspective. Volume II.* UN Document Repository. https://unstats.un.org/unsd/ccsa/documents/covid19-report-ccsa_vol2.pdf

United Nations Environment Programme (UNEP). (2021). *How COVID-19 is changing the world: A statistical perspective. Volume III.* UN Document Repository. https://unstats.un.org/unsd/ccsa/documents/covid19-report-ccsa_vol3.pdf

Wallman, K. K. (1993). Enhancing statistical literacy: Enriching our society. *Journal of the American Statistical Association, 88*(421), 1–8.

Watson, J. M. (1997). Assessing statistical thinking using the media. In I. Gal & J. B. Garfield (Eds.), *The assessment challenge in statistics education* (pp. 107–121). IOS Press and The International Statistical Institute.

Watson, J. M. (2011). Foundations for improving statistical literacy. *Statistics Journal of the IAOS*, *27*(3–4), 197–204. https://doi.org/10.3233/SJI-2011-0728

Watson, J. M., & Callingham, R. (2003). Statistical literacy: A complex hierarchical construct. *Statistics Education Research Journal*, *2*(2), 3–46.

Watson, J. M., & Callingham, R. (2020). COVID-19 and the need for statistical literacy. *Australian Mathematics Education Journal*, *2*(2), 16–21.

Wild, C. J., & Pfannkuch, M. (1999). Statistical thinking in empirical enquiry. *International Statistical Review*, *67*(3), 223–265. https://doi.org/10.1111/j.1751-5823.1999.tb00442.x

Wild, C. J., Utts, J. M., & Horton, N. J. (2018). What is statistics? In D. Ben-Zvi, K. Makar, & J. Garfield (Eds.), *International handbook of research in statistics education* (pp. 79– 96). Springer. https://doi.org/10.1007/978-3-319-66195-7_1

Yildirim, T., Shuman, L., & Besterfield, M. (2010). Model-eliciting activities: Assessing engineering student problem solving and skill integration. *International Journal of Engineering Education*, *26*(4), 831–845.

Yoon, C., Dreyfus, T., & Thomas, M. (2010). How high is the tramping track? Mathematising and applying in a calculus model-eliciting activity. *Mathematics Education Research Journal*, *22*(2), 141–157. https://doi.org/10.1007/BF03217571

CHAPTER 8

Fostering Student's Mathematical Literacy through Mathematical Modelling Tasks

Peter Frejd, Jonas Bergman Ärlebäck and Pauline Vos

Abstract

In this chapter, we compare mathematical modelling and mathematical literacy, and explore mathematical modelling tasks as a means for fostering students' mathematical literacy. We draw on our own research on three different classroom-based mathematical modelling tasks in different contexts that yielded rich and varied student activities. We analyse the tasks in light of five components of mathematical literacy: developing human capital, maintaining cultural identity, pursuing social change, creating environmental awareness, and evaluating mathematical applications. These five components can be used to enact or evaluate teaching that fosters students′ mathematical abilities to participate effectively and democratically in society in a productive and critical manner. We use these five components to evaluate how the three modelling tasks can foster mathematical literacy. The results show that all components came to the fore in at least one of the three analysed modelling tasks. More generally, our approach to evaluating mathematical modelling tasks serves as an illustration for teacher educators and teachers concerning how they can analyse modelling tasks to foster mathematical literacy.

Keywords

digital simulations – fostering – hands-on activities – mathematical modelling – mathematical literacy – Sankey diagrams

1 Introduction

"Why should we learn this?" is a recurring question asked by students in mathematics classrooms when they encounter mathematics that is disconnected from real-life (Brown et al., 2008; Hernandez-Martinez & Vos, 2018). Teachers say that they would like to be better at explaining the usefulness of mathematics to their students, but that they are hindered by a lack of time, resources,

 | DOI:10.1163/9789004710962_008

or knowledge about such usefulness (Gainsburg, 2008). Moreover, knowledge about its usefulness is rarely the focus of examinations (Prieto-Rodriguez, 2016). Nevertheless, when students recognize usefulness in what they learn, they are more interested and confident (Sheldrake et al., 2017). They spend more time on tasks, have a better attitude towards the content and achieve higher grades (Grootenboer & Hemmings, 2007).

There have been different approaches to better connecting school mathematics and life beyond the school gates. One approach has been to include word problems in mathematics classrooms, which are tasks describing realistic situations that end with a mathematical question. Verschaffel et al. (2000) give several examples, such as tasks about mothers being exactly three times the age of their daughter that end with a question about the age of the father three years ago. However, such tasks are clearly constructed by someone who already knows the ages, and thus is not in need of an answer. Moreover, in everyday life, one would rather ask one of the family members for the father's age and not set up algebraic equations. Although some people prefer these types of tasks for their puzzle characteristics, the detrimental effect of word problems is that they confirm students' impression of the uselessness of mathematics (Verschaffel et al., 2000).

Another approach to strengthening the connection between school mathematics and real life has been to call for the inclusion of mathematical literacy in mathematics curricula, as explained in other chapters of this book. However, mathematical literacy is often described as a goal by educational and governmental authorities (OECD, ministries of education, school leaders). This leaves mathematics teachers with insecurities concerning how this goal can be addressed in everyday mathematics classrooms and the daily teaching and learning of mathematics (Bolstad, 2023).

In this chapter, however, we focus on a third approach aimed at developing students' competency in using mathematics to solve real-life problems. This approach emerged in the 1970s, when mathematics teachers, often individually, introduced mathematical modelling tasks to their students (Niss & Blum, 2020). The community of teachers of mathematical modelling grew, and an international research field gradually took form, which now yields a growing body of research as well as many modelling tasks for use in mathematics education. Mathematical modelling tasks for middle- and high schools are bundled into topical books in various languages (e.g., Erbaş et al., 2016; Gould et al., 2012; LEMA, 2009; Maaß, 2008), and it is the same for tertiary STEM education (Science, Engineering, Technology, Mathematics).

In this chapter, we report a research-informed analysis of three modelling tasks used in middle school, upper secondary school and teacher education

(Ärlebäck et al., 2021; Ärlebäck & Frejd, 2020; Vos & Frejd, 2022). We present these because of their novelty in the continuously growing collection of mathematical modelling tasks. Our aim is to investigate the extent to which these three tasks can be used to foster mathematical literacy. This analysis can also serve as an approach so that, if followed, teachers can evaluate other mathematical modelling tasks regarding their suitability for fostering students' mathematical literacy.

Before we analyse the three modelling tasks, we set the scene by briefly describing *mathematical literacy, mathematical modelling* and *comparing the two*, after which we will explain the term *fostering* as an important approach and integral part of teaching in supporting students to develop both mathematical modelling competency and mathematical literacy.

2 A Short Review of Mathematical Literacy

The term mathematical literacy has already been explained in greater detail in previous chapters of this book. We acknowledge that mathematical literacy is complex, multi-faceted and includes the use of mathematical knowledge (concepts, skills, problem solving and estimation strategies), positive dispositions, and tools (representational, physical and/or digital) in a range of personal-, social-, and work contexts, all of which contribute to informed and critical citizenship. More practically for this chapter, we will use a short working definition of mathematical literacy, to be understood as those mathematical competencies the general citizen needs to participate effectively and democratically in society and to contribute to it in a productive and critical manner (Frejd & Geiger, 2017).

To analyse the appropriateness of applying the three modelling tasks for developing mathematical literacy later in this chapter, we will use the *components* of mathematical literacy as described by Jablonka (2003) and listed below.

1. *Developing human capital*

   Mathematical literacy for developing human capital concerns the intention that all citizens should be able to use school mathematics to improve their daily situations, to get 'good' jobs and to solve problems in their social and work environments.

2. *Maintaining cultural identity*

   Mathematical literacy for maintaining cultural identity is connected to ethnomathematics (D'Ambrosio, 1985) and refers to the application of informal mathematical knowledge in cultural heritage practices, such as weaving, knitting (Chapman, 2020) or fishing (Fyhn & Robertsen, 2020).

3. *Pursuing social change*

 Mathematical literacy for pursuing social change is related to an ability to make impact on political debates as an informed and critical citizen by using arguments based on mathematical reasoning, in particular about issues concerning social equity (poverty, emigration, addictions, obesity, etc.).

4. *Creating environmental awareness*

 Mathematical literacy for creating environmental awareness relates to the use of mathematics for understanding problems about energy production and consumption, ecosystems, and other aspects of our natural environment. This component includes abilities for interdisciplinary work, teamwork, and the use of technology.

5. *Evaluating mathematical applications*

 Mathematical literacy for evaluating mathematical applications relates to citizens being reflective and critical of the use of mathematics in terms of being sceptical of its apparent objectivity and its limitations.

For a more comprehensive description of the five components of mathematical literacy we recommend Jablonka (2003). We now turn to a short description of mathematical modelling, after which we will briefly compare mathematical literacy to mathematical modelling.

3 A Short Review of Mathematical Modelling

In this chapter, we take mathematical modelling to be a generic real-world problem-solving process where the use of mathematics is an integral and essential component (Niss & Blum, 2020). One perspective frequently adopted is that of mathematical modelling as a cyclic process addressed in research literature as *the modelling cycle* (Niss & Blum, 2020; Geiger & Frejd, 2015). The modelling cycle can be organized and depicted in many ways with different foci (Borromeo Ferri, 2006; Jablonka, 1996; Perrenet & Zwaneveld, 2012). Figure 8.1 displays an example that shows two domains – reality and mathematics – as well as a set of transitions (Blum, 2015).

Figure 8.1 shows that a modeller makes transitions between reality and mathematics and back to reality in a cyclic process by making the problem context mathematizable and interpreting and validating the mathematical results.

Modelling cycles, as depicted in Figure 8.1, are a means to explain how modelling works, that is – how a modeller starts with a real-life problem, uses

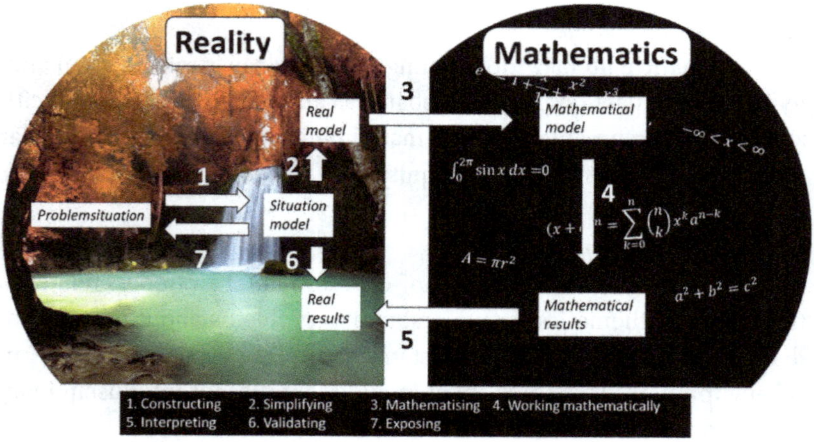

FIGURE 8.1 Modelling cycle inspired by Blum (2015)

mathematics and ends with presenting (exposing) the outcome (real results) to others, usually to the one who came up with the problem. However, a modelling cycle neither gives a recipe to follow, nor does it reveal that modelling processes are often much more chaotic, whereby modellers move back and forth between the different transitions (Borromeo Ferri, 2006; Ärlebäck, 2009). Thus, the modelling cycle is just *a model* to be used as a tool for researchers and teachers to recognize, comprehend, analyse and explain different aspects in modelling (Vos & Frejd, 2022).

4 Comparing Mathematical Literacy and Mathematical Modelling

In this subsection, we compare mathematical literacy and the perspective of mathematical modelling as illustrated in Figure 8.1. We initially note that 'literacy' relates to competency (of people), whereas modelling relates to 'activities' (by people). An associated term to mathematical modelling, and more compatible to mathematical literacy, as used in the literature, is 'mathematical modelling competency', which means "being able to autonomously and insightfully carry through all aspects of a mathematical modelling process in a certain context" (Blomhøj & Jensen, 2003, p. 126). However, in this definition the term mathematical modelling is so central that we continue the comparison of mathematical literacy and mathematical modelling, ignoring differences between competency and activities.

Both mathematical literacy and mathematical modelling have in common that they relate mathematics to practices outside the discipline of mathematics. For mathematical literacy and mathematical modelling, mathematics is not celebrated for its purity, conciseness, immateriality, generality, or abstractness, but both emphasise the service role of mathematics. So, rather than perceiving mathematical concepts, and algorithms as objects in themselves, these are now perceived as tools in a mathematical toolbox to assist in private and professional situations.

Also, mathematical literacy and mathematical modelling share the strong research evidence of their positive and lasting effects on students' overall mathematics achievement and attitudes (e.g., Boaler, 1998; Holenstein et al., 2020; Sokolowski, 2015). In the same way as students improve their knowledge of a foreign language through *conversations* (using the language flexibly in new situations), and even improve their knowledge of grammar, the flexible use of mathematical concepts and algorithms improves students' understanding of these mathematical objects, their appreciation of these and their confidence in using them in unfamiliar situations.

When looking at differences, we see that mathematical literacy takes a perspective that starts from the societal (macro) level, then moves towards the individual (micro) level. In starting from society, the perspective of mathematical literacy specifies the mathematical abilities needed for individual citizens to participate in and contribute to society. Central to this perspective are the verbs to *contribute* and to *participate*, but these verbs do not explicitly implicate a certain ability in problem solving. For instance, from the mathematical literacy perspective, a mathematically literate citizen should be able to read and make sense of line graphs published in the general news media, so they can look-up the weather forecast or see that COVID-19 infections are rising. This sense-making of line graphs assists the citizen in participating in society. However, such reading of graphs is not, in itself, mathematical modelling, since the reading citizen may not contribute to the solving of a problem. In case the citizen wants to solve certain problems, such as "what clothes will I wear today?" or "is it safe to mingle socially?", the use of the line graphs could be part of a mathematical modelling activity.

There is another difference between the perspectives of mathematical literacy and mathematical modelling. The mathematical modelling perspective does not per se state at what level of sophistication the modelling takes place. The modelling can be carried out by students, citizens, or professional modellers. The professional modelling work by the latter category is excluded from the mathematical literacy perspective, since the solving of problems through advanced mathematical techniques is generally not done by the average citizen for participating in, and contributing to, society.

Another contrast is that the mathematical modelling perspective does not state relations between the modelling activity and society. First, there is no indication of whether the modeller will participate in and contribute to society; the modeller could theoretically also perform the modelling activity in social isolation and without social purpose. Second, the mathematical modelling perspective does not state where the real-world problem must come from; they could emerge from textbooks or from someone's imagination and be in no way related to the modeller's participation and contribution to society, let alone the five components of mathematical literacy as set out by Jablonka (2003), such as developing critical citizenship, cultural identity, and environmental awareness.

Thus, we acknowledge that the perspectives of mathematical literacy and mathematical modelling have differences. Nevertheless, we noted commonalities, which enable teachers to design, implement, and evaluate learning activities based on existing modelling tasks that can simultaneously support students in developing mathematical literacy.

We next discuss what teachers can do to support students in developing mathematical literacy and we use the term fostering for that.

5 Fostering

Being mathematically literate and a competent mathematical modeller have in common that neither directly results from traditional teaching of (pure) mathematics. There is ample evidence that the traditional teaching of pure mathematics insufficiently prepares students to use and apply learned mathematical concepts and algorithms outside the mathematics classroom, such as in daily life or in other subjects; this problem is known as the 'transfer problem' (Lerman, 2012). Within the research paradigm of mathematical modelling, an established teaching principle is that "[m]odelling competency is developed through the practice of modelling" (Blomhøj & Hoff Kjeldsen, 2006, p. 166), meaning the students should be provided with opportunities to work with 'realistic' modelling problems in their education. Metaphorically, learning mathematical modelling is like learning to swim: one cannot expect that swimming will be learned after, in isolation, first exercising breathing, kicking with one's legs and stroking with one's arms on the beach – at some point, one needs to do it all at the same time in water. We acknowledge that there is a clear difference between knowing in theory and knowing in practice, and we argue that the teaching principle above that emphasizes a nurturing environment in which students experience the solving of realistic problems is also adequate for the teaching and learning of mathematical literacy.

We will use the notion of *fostering*, which is different from direct instruction, whereby a teacher shows and explains how tasks are to be done (Frejd & Vos, 2022). The term fostering is frequently used, but rarely defined, in research literature. In the Cambridge Dictionary (n.d.), fostering is to encourage the development or growth of ideas or feelings. This meaning is associated with motivation and facilitating something. Thus, fostering relates to encouraging and helping someone to achieve *something*, through nurturing, socialization, etc. Such processes are usually time consuming and goal directed, which requires trust and interaction between the individual(s) doing the fostering and the individual(s) being fostered. This interaction can be verbal or non-verbal. For example, parents can foster their children by using body language, eye contact, gestures, etc., to facilitate a desired behaviour.

In research literature, the notion of fostering is linked to aspects such as creativity (Munakata & Vaidya, 2013) and collaboration (Chih, 2021). Fostering in this context "indicates an indirect, exploratory approach" (Kokkinos, 2009, p. 4). In an educational setting, we argue that fostering is a collaborative process that is more student-centred, more informal, more experiential, and more inspiring compared to traditional teaching. Unlike mathematical objects, concepts, or algorithms, neither mathematical modelling nor mathematical literacy are likely to be learned through direct instruction. However, tasks can be designed that foster mathematical literacy, which should recognize learners as active participants in their own learning (Haara, 2018) and provide them with opportunities to use their own 'voice and choice'. To accomplish such goals requires that the teacher and students together create a classroom culture of creativity and patience with an openness for critical and supportive discussions. The tasks should allow the students to receive ample time to be creative, to struggle, and to reflect about the outcomes. Another component in designing fostering activities is to use the students' own intrinsic motivation to learn, based on their interests, that will facilitate students in developing the agency and ownership of their own learning (Doerr & Ärlebäck, 2015). Thus, the design of the tasks should engage the students in discussing issues important to them beyond the mathematical classroom. In addition, an integrated part of the activities should also explicitly foster collaboration abilities, and students should become familiar with what effective collaboration is, as well as what it looks, feels, and sounds like.

With this chapter, we want to illustrate, with three different examples of modelling tasks implemented in classroom settings, how they in different ways impact the fostering of the use of mathematics to support participatory, functional, and critical citizenship (Bolstad, 2023).

6 The Description and Analysis of the Three Modelling Tasks

In the following sub-sections, we will showcase three modelling tasks (see Table 8.1) to evaluate the extent to which they foster different aspects of mathematical literacy. To explore how the three modelling tasks can foster students' mathematical literacy, we first briefly describe their instructions and then analyse them and their implementations using the five components of mathematical literacy by Jablonka (2003). The three researchers independently analysed the tasks and their implementations by matching Jablonka's (2003) components with information from the implementations of the tasks, and all researchers developed notes with arguments for characterizing or not characterizing the components. These notes were thereafter discussed between the researchers to gain consensus.

TABLE 8.1 The three modelling tasks

Task	Educational level	Aim of task	References
Polling	Teacher education	Developing pre-service teachers' understanding of how to construct sampling models for different population sizes to make inferences about the outcome of an election	(Ärlebäck & Frejd, 2021; Ärlebäck et al., 2021)
Pandemic	Upper secondary education	Developing students' critical understanding of the role of mathematical models in predicting and handling the spread of infection diseases	(Ärlebäck & Frejd, 2020; Frejd & Ärlebäck, 2017)
Sankey-Diagram	Middle school	Developing students to learn to read, interpret and create so-called Sankey diagrams about societal issues (e.g., population growth, electricity usage of lightbulbs)	(Vos & Frejd, 2020, 2022; Vos & Omobude, 2021)

To clarify our language of description, *a task* is a set of guidelines and questions that are designed *before* a teacher enters the classroom. When the teacher *enters* the classroom with the tasks, these are then implemented (or 'enacted'), that is – students start working with them in *an activity*.

6.1 *The Modelling Task Polling*

The task Polling was designed as an interdisciplinary activity connecting mathematics and the social sciences in the context of elections and opinion polls. The task aimed at fostering students' ideas about sampling strategies and the role sampling size plays in opinion-polls when devising a model to calculate a reliable inference about a population of a given size. The two-hour modelling task began with the introduction and contextualising of the task using a presentation providing information about the notion of opinion(s) and some history about opinion polls. The design of the task followed the principles of the models and the modelling perspective on teaching and learning (Lesh & Doerr, 2003) and included hands-on material that emulated the situation investigated: seven containers that contained a predefined number of beads of different colours (see Figure 8.2). Each bead represented an individual in the population with a possibility to vote (or express an opinion), and the different colours of the beads represented the different categories of the attribute being examined, such as casting a ballot for a particular candidate or party. The colour distribution in six of the containers was unknown to the students, whereas one container with 20 beads was open for inspection. The latter container was provided to facilitate students' engagement with the activity.

After the introduction of the task the students worked in groups of two to four students on the following tasks (Ärlebäck & Frejd, 2021, p. 105):

> Develop a sampling model to investigate the following two questions:
>
> Q1. How many different colours of beads are there in a given container containing N beads?
>
> Q2. What is the distribution of coloured beads in a given container containing N beads?
>
> The sampling model you develop should be applicable to any container containing an arbitrary (but known) number of beads. Test your sampling model on the six provided containers containing N different coloured beads (N = 20, 400, 10,000; two containers of each size) with an unknown distribution and reflect on your sampling model. Document your work on the provided worksheet.

FIGURE 8.2 The six containers of beads representing the unknown populations (N = 20, 400, 10,000; two containers of each quantity – one uniformly distributed and one skewed)

The students were also provided with a closed envelope with data on the actual bead distributions of the six containers to open. They then used their developed model and made the requested inferences about the unknown distributions of the six populations. The intended idea behind providing the actual distributions, for reference after sampling and drawing inferences, is that they could serve as a reference in discussing and writing about the strengths and limitations of their developed sampling models.

Note that there is a difference between the two tasks Q1 and Q2. The first question asks the students for the number of different colours in a population, which is to make an inference about a single variable, whereas the second question asks for a multi-valued statistic in terms of an inference with respect to the distribution of colours in a population. In addition, the groups of students were provided with a sampling device consisting of a plastic pipe of suitable diameter cut in half lengthwise with closed ends, to be used to facilitate the actual sampling.

6.2 *A Literacy Perspective on the Modelling Task Polling*

The task Polling provides a context of opinion polls or elections, which are contexts a citizen will experience in their lives in one way or another. To gain knowledge about how to establish reliable samples and make valid inferences is a central ability for an individual to critically reflect on and interpret quantitative information provided in media (Gal, 2004). The component of mathematical literacy called *developing human capital* (Jablonka, 2003) aligned with the aim of the task, where the use of (formal) mathematics is in focus in a simulated realistic problem of developing accurate inferences based on collected samples. In addition, the task includes working in groups and asks students to evaluate the sampling models they developed, which corresponds to the aspect of teamwork in the component of *creating environmental awareness* as well as the component of *evaluating mathematical applications*. However, the task was designed as a modelling activity to elicit students' knowledge of sampling models as a starting point for a more robust discussion of the mathematics used in formal settings. From the implementation of the task, we found that the students also developed their models with no or few connections to the standard formulas used in statistics, which is connected to the informal use of mathematics as a component of *maintaining cultural identity*. Nevertheless, the task analysed from a perspective of mathematical literacy tends to have several shortcomings. The students quickly disregarded the context of opinion polls and spoke only about the colours of the beads and number of beads. In addition, the developed sampling models were only evaluated on the bases of the populations of beads, and the pseudo-realistic issue about interpreting and evaluating elections or opinion polls was not actually discussed during or after the activity.

In terms of fostering mathematical literacy, the task facilitated students to develop their own (group) cognitive strategies for creating mathematical models. However, the situation about opinion polls and elections was already translated into mathematical tasks in the instructions, and the activity was prepared with equipment of containers with beads, which tended to foster more specific mathematical abilities rather than mathematical literacy. During the implementation of the task, the students in the different groups worked on the tasks and all groups developed some type of formula, which shows an example of how the power of social norms directs the attention of fostering due to the prevailing didactical contract (Brousseau, 2002) and socio mathematical norms (Yackel & Cobb, 1996). The outcome of what is fostered is not always transparent in students' written solutions. For example, the students' developed models presented in their written responses, and their reasoning during the solution process shows two sides of the same coin. Where the written responses only provided a 'short' description of the solution, with a formula that had little or no reflection about validity or reliability, the video-recorded discussions of the students' solution processes showed more nuanced and reflective reasoning by the students (Ärlebäck et al., 2021). For example, the video recordings show that a single yellow bead that was included in the sample in the seventh container influenced the students' thinking and their developed sampling models by engaging the students in self-evaluating during their modelling process.

In summary, fostering mathematical literacy by implementing the Polling task stresses a (pseudo)realistic context and leads to a group activity focusing on developing and validating mathematical models using formal and informal reasoning. This highlights that even if the aim of the task focused on elections and opinion polls, the actual implementation showed that the traditional classroom norms and expectations that focus on the learning of (pure) mathematics was adopted rather than a fostering of mathematical literacy.

6.3 *The Modelling Task Pandemic*

The second modelling task, Pandemic, was designed (and first implemented) for upper secondary students before the Covid-19 outbreak. The aim of the task was to foster students' understanding of how mathematical models can be used to predict and make sense of exponential phenomena in the context of the spread of infection diseases (Ärlebäck & Frejd, 2020; Frejd & Ärlebäck, 2017). By using a commercial (but free) mobile game app for smartphones and tablets, students participating in the activity were provided with a context and a digital simulation of a pandemic outbreak that they explored using mathematics. The games used were *Plague* and *Infection*, which are similar in terms of their design, controls, functions, etc. The goal in both games is to wipe out the world population as fast as possible by making various in-game-decisions affecting the nature and speed of how a virus or bacteria evolves. The simulation of the pandemic begins after the player decides on the level of difficulty, the name of the virus and the country where patient zero is infected. During the pandemic simulation the player receives data such as the number of infected and the number of deceased at any given time. In addition, the player sees a news banner and a calendar at the top of the screen, which makes it possible to track elapsed time and dates. Information about the progress of a cure that is being developed, as well as the number of so-called DNA points collected, is also visible to the player. The DNA points is the game's currency gathered in various ways during the game, such as touching circular symbols in the world map (see Figure 8.3)

By touching the *Disease* button (Figure 8.3a), a new screen appears (Figure 8.3b) where collected DNA points can be used to evolve and mutate the virus, which affects its infectivity, severity, and lethality, as well as the virus' method of transmission, the symptoms induced, and its resistance to drug treatments. Touching the *World* button (Figure 8.3a) also generates a new screen (Figure 8.3c), containing data about the spread of the infection, the progress of the cure, and statistics on the numbers of infected and deceased.

FIGURE 8.3 Screen shots of (a) the default Map tab; (b) the Disease tab for evolving the disease; and (c) the World tab with a statistical summary and information about the progress of the cure being developed

The task aimed at addressing the following learning goals in the Swedish mathematics curriculum: "to use digital tools, to use mathematics for exploring realistic scenarios, to interpret and use different representations of data such as tables and graphs, and to investigate the role of exponential functions in a given context" (Skolverket, 2012). The introduction and contextualizing of the task included historical facts about pandemics and how pandemics evolve. The instructions to the students were as follows:

> Play the game at the easiest level; take data points in terms of the number of infected and dead in the world every other week; record the data in an

> Excel spreadsheet; play until you win or lose; and draw a graph of how the number of infected and dead evolved as the game progressed.

The instructions in addition included ten questions for students to answer. Examples of questions are the following:

> Draw rough sketches of graphs and briefly describe the course of the pandemic outbreak. Is the number of infections decreasing? (And if so, when and why?) Is there a relationship between the graphs of the numbers of infected and dead people? When is a world citizen at the greatest risk of contracting the disease?

6.4 *A Literacy Perspective of the Modelling Task Pandemic*

Similar to the Polling task, the Pandemic task has a clear connection to real life issues, as it is about the spread of infectious diseases. The research studies were conducted before the Covid-19 pandemic, and the students were not as familiar with the phenomena and issues as they may be now. To understand how the spread and effects of diseases evolve and to interpret graphical representations of the number of infected and deceased people during a pandemic is an ability that is needed, for example during the height of the Covid-19 pandemic and the concurrent media flow. In other words, the task emphasises the component *developing human capital*, that is, using formal mathematics to explain and predict real-life issues using different types of representations. Compared to the Polling task, the Pandemic task did not remove the authentic aspect by using hands on material (beads with different colours) to simulate people. Instead, the game apps had features of a more virtual realistic simulation context in terms of news reports illustrated by a world map and statistics concerning how the spread and effects of the disease evolved. This use of a digital tool contributes to developing human capital. To facilitate data collection and the discussion of answers, the students were set up to work in pairs, and such group work aligns with the components of *creating environmental awareness*. The component *evaluating mathematical applications* of mathematical literacy is found directly in the instructions where the students were instructed to evaluate and explain their models. The implementation of the task, however, indicated that the task did not foster students to use formal mathematics in order to support their claims. Rather, they enacted the language provided by the game context. Hence, it can be seen as an aspect of the component of *maintaining cultural identity*. The game apps were also adaptable due to options made by the player, who could thus make an impact on the simulation and therefore the outcome of the game. In this sense, the gamer could be seen as *pursuing social change* in the virtual realistic simulation context, by manipulating the behaviour and features of the virus in the game.

The Pandemic task aims to foster an exploration of the "black box" of the game app simulation, namely, the mathematical models developed through the game. The activity also includes several digital tools: both the simulation apps and Excel for recording and visualising the data collected. Thus, the task can be seen as aiming to foster different aspects of "tools use". Indeed, the game apps provide a realistic virtual context and generate the data to explore, while Excel transforms the data collected into a diagram with a function describing how the number of infected people changed with time; it also expressed the number of deceased as a function of time. The game apps were also used in the evaluation process by the students. They could verify that the function graph they had developed looked accurate by comparing and mapping the shape of the graph to the modifications they had induced on the virus at certain points in time. The features of the generated graphs and the relation to the collected data could possibly have fostered and increased students' understanding of how the graphs were used on a daily basis for illustrating the spread of the ongoing Covid-19 pandemic. However, the task tended to preserve social norms about gaming. Boys play video games to a larger extent than girls, and there were statistically significant differences in the attitudes between female and male participants, where the male players enjoyed the task more than the female players (Frejd & Ärlebäck, 2017).

In summary, the Pandemic task can be seen to foster several aspects of Jablonka's (2003) components of mathematical literacy, in particular by focusing on the use of digital tools and the understanding of the underlying mathematical structures hidden in artifacts. In addition, the game context enabled the students to investigate social change within a virtual world. However, as implemented, the task did not foster the use of formal mathematics as a tool for reasoning; instead, the game's play culture prevailed and social norms about attitudes towards these types of games were preserved.

6.5 *The Modelling Task Sankey Diagrams*

The modelling task Sankey Diagrams was carried out with 8th grade students in a collaboration between a Norwegian middle school and researchers of mathematical modelling education. After the students had engaged in the activity, the same students participated in a waste sorting project, which we also describe because it was affected by the Sankey diagrams task. A second reason for describing the waste sorting project is that it was never intended as a mathematical modelling activity, but rather spontaneously turned out to be just that. Although interrelated, the two activities were independent. Thus, teachers can choose to implement only one of the two activities.

The task Sankey Diagrams was initiated after an excursion by the students to a local enterprise, a sewage plant, and the students tried to describe the sewage process by drawing sketchy flow diagrams. With a goal to use a more formal mathematical description for such complex industrial processes, the students were offered a mathematical tool called Sankey diagrams. These types of diagrams are mathematical models to visualise the flows of people, money, fluids, and so forth. Figure 8.4 shows three Sankey diagrams displaying the conversion of electricity in lightbulbs into light and heat created by students. In Sankey diagrams, the width of a flow represents the flow's quantity. With such diagrams, one can show how a whole has parts, and these parts can again be divided into sub-parts.

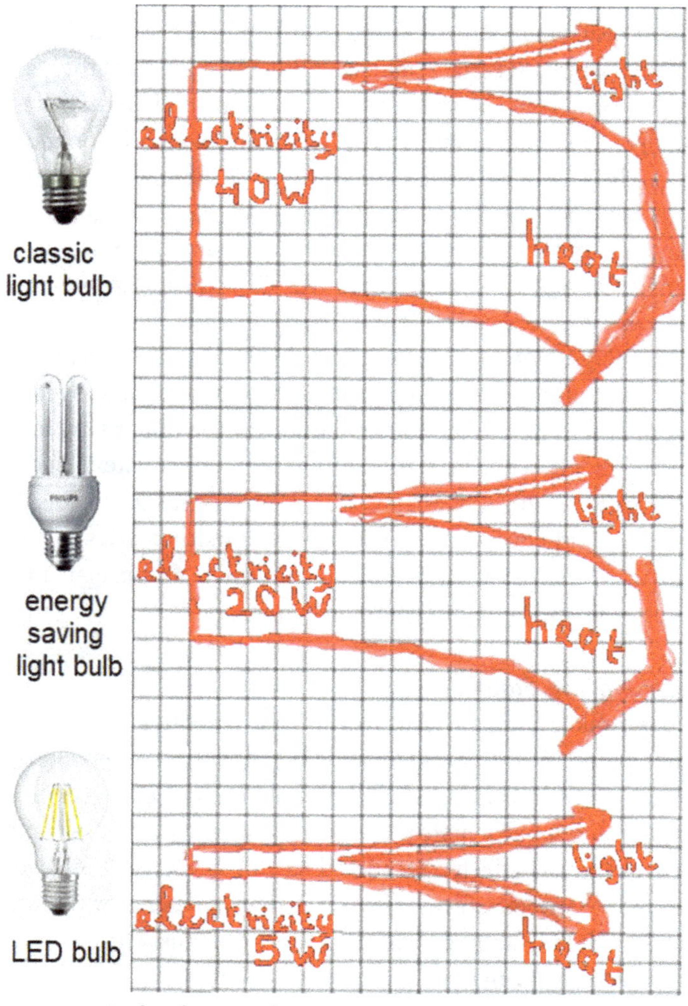

FIGURE 8.4 Sankey diagrams showing energy conversion in lightbulbs

In a 90-minute lesson, groups of two or three students worked through an 8-page booklet (Vos, 2019) that aimed at developing an understanding of Sankey diagrams. The modelling tasks about developing and analysing mathematical models were, for example, what are reasons for Norway's population growth? Or, what are the differences in energy usage between classical light bulbs and energy saving bulbs? In the case of population growth, students were given a ready-made Sankey diagram with flows representing newborns, deceased, immigrants, and so forth. In the case of the light bulbs, students were given data to draw Sankey diagrams for the lightbulbs (see Figure 8.4). At the end of the lesson, students were generally able to read and make sense of Sankey diagrams, and adapt and refine them, as well as by hand draw a simple Sankey diagram about an issue of their interest (Vos & Frejd, 2020, 2021; Vos & Omobude, 2021).

Two months after working on the Sankey Diagrams task, the same class engaged in a waste sorting project, which was thought by the teacher to be unrelated to the subject of mathematics. The students were asked by the local recycling enterprise to study how citizens could improve their waste recycling. The students were to present their findings after three weeks to the leaders from this enterprise. The students independently, but supported by their teacher, decided to re-word the problem into the question "How much waste in the black bins (non-recyclable waste) at our school should be in some of the other bins?" This question required them to use mathematics to analyse the unsorted waste. With respect to this insight, the mathematics teacher commented that, in the design of projects, teachers do not need to explicitly aim for mathematics, since one can trust that mathematics will appear, because "mathematics is everywhere".

The students worked in groups and overturned the black bins from the school yard making a big pile of waste. They sorted it into smaller piles of recyclable materials, such as 'organics & food', 'paper & cartons', 'metal & glass', and 'plastics, as well as one pile of 'unrecyclable waste'. The students weighed the sorted piles in grams. To their dismay, they discovered that approximately 90% of the waste in the black bins at the school yard was recyclable. In preparing for their presentation to the leaders of the recycling enterprise, the students had independently (but encouraged by their teacher) found a web-based drawing program for producing Sankey diagrams. They figured out and taught each other how to use it. Figure 8.5 displays one of their diagrams, showing that in a pile of 967 g waste in the black bins from the school yard, there was only 91 g unrecyclable material ('Restavfall').

Then, the students discussed with their teacher that they wanted to request more recycling bins in the school yard from the school authorities. The teacher suggested that they should send a delegation rather than the whole class going.

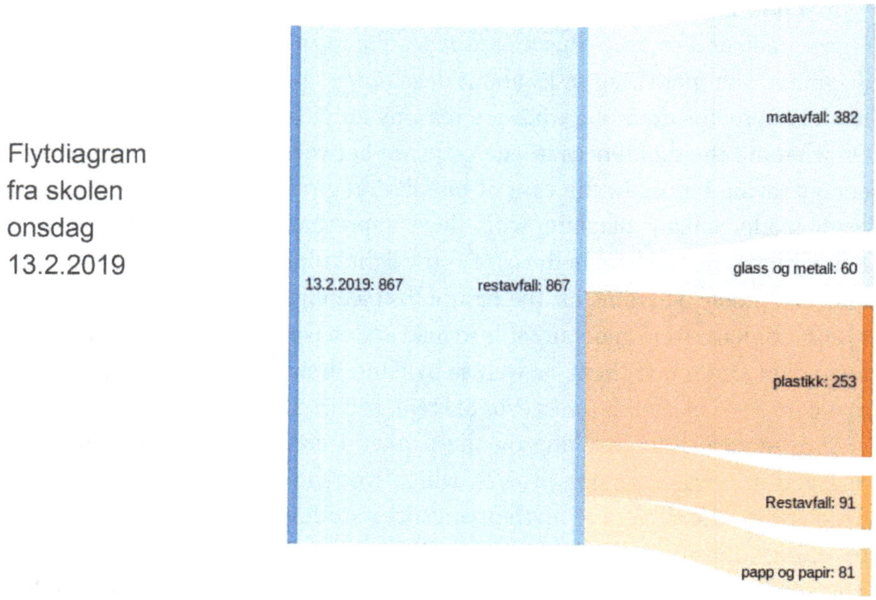

FIGURE 8.5 Sankey diagram designed by the students in the Project Waste Sorting. Text on the left: "Flow diagram from the school Wednesday 13th February 2019". Text in the middle: "13th February 2019: 867 unrecyclable waste: 867". Text on the right: "food waste: 382, glass and metal: 60, plastics: 253, unrecyclable waste: 91, carton and paper: 81"

The class then selected four students and jointly prepared a presentation. To convince the headteacher, they used their Sankey diagrams. Their message was successful, and he promised to buy new recycling bins for the school yard. A week later, the students presented their findings to the leaders of the recycling enterprise, in which they again included their Sankey diagrams, and how they had entered into a dialogue with the school authorities about acquiring new recycling bins.

6.6 *A Literacy Perspective on the Task Sankey Diagrams*

The Sankey Diagrams task was implemented using a booklet that described societal issues such as population growth and energy conversion in different lightbulbs. It provided contexts in a similar vein to the Polling- and Pandemic tasks, but in this case all data were authentic. The mathematical models used were the Sankey diagrams, and the booklet offered these ready made for interpretation and sense-making tasks. Only at the end of the booklet were students asked to draw their own Sankey diagram to model a phenomenon of their own interest. Thus, most parts of the booklet did not offer mathematical modelling tasks, whereby the students would 'run' through a full modelling cycle (see Figure 8.1). Instead, the students would mostly 'work mathematically' and 'interpret', and only towards the end of the booklet would they engage in 'mathematizing'.

When analysing the task Sankey Diagram in light of Jablonka's (2003) five components, the task emphasises the category *developing human capital* by using mathematical representations to visualize social phenomena. The energy conversion task, and later the waste sorting project, were clearly also aiming for *environmental awareness*, since both examine the use of resources, saving energy and re-using materials. The components of *evaluating mathematical applications* and *maintaining cultural identity* were, however, not addressed in this task.

A striking feature in the waste sorting activity is that it enabled the students to *pursue social change*. The students did not just construct a mathematical model for communication, they deliberately used their developed model to convince the headteacher at the school to change the social environment by providing more recycling bins for the school yard. Thus, the mathematical model was not only a tool to understand a problem, but also a communication tool to persuade the authorities.

When looking at the fostering role of the teacher, we can make two observations. First, in the task Sankey Diagrams, the teacher stood at the side while the students worked in groups with the booklets. Booklets, as well as worksheets, are tools that teachers can use to offer students a certain amount of freedom and autonomy. In the case of the Sankey Diagrams task, each group of students had only one booklet and had to jointly write their answers, which was another way to encourage their collaboration. In the subsequent waste sorting activity, we saw that the teacher allowed students much agency by, for instance, allowing them to change the initial problem into a researchable question, and encouraging them to send a delegation to the school authorities. This teaching approach created student ownership and gave students control of their own learning.

In summary, the connection of mathematical literacy and mathematical modelling really comes to the fore in the Sankey Diagrams task and the follow-up waste sorting project. The ultimate aim in both mathematical modelling and mathematical literacy is to use mathematics in such a way that students can impact and change their social and environmental situations in and beyond school 'for the better'.

7 Discussion and Conclusion

In this chapter we applied Jablonka's (2003) five *components* of mathematical literacy (developing human capital, maintaining cultural identity, pursuing social change, creating environmental awareness, and evaluating mathematical applications) for the analysis and discussion of three different mathematical modelling tasks (Polling, Pandemic and Sankey diagram). We have illustrated that it is possible to foster mathematical literacy in connection with mathematical modelling and highlighted that Jablonka's (2003) components come to the fore in the three analysed tasks in different ways. All three tasks emphasised working together in groups to solve problems in society involving contexts beyond the current tradition of school mathematics, thereby producing relevance for students to learn and apply mathematics in realistic situations (Hernandez-Martinez & Vos, 2018). Developing human capital in terms of applying formal mathematics was the goal in all three tasks. However, during the implementation of both the Polling- and the Pandemic tasks students tended to adopt homegrown strategies (informal mathematics). Their development and critical analysis of the mathematical models were affected by the social and cultural contexts in which the activities took place, demonstrating the aspect of maintaining cultural identity. Some elements of the purposeful design of the tasks, such as the picking tool for developing samples, and the students′ lack of experience with similar hands-on activities, were some issues impacting implementation in the Polling task. Exploring mathematical applications in terms of mathematical models was an explicit question in the Pandemic task, and here the app functioned as a black box generating the data to study, and the students were unaware of the underlying mathematics. Hence, they could not evaluate their models adequately. In the Polling task, however, the students were explicitly asked to evaluate their answers based on "knowing" the correct distributions of the populations. In real-life situations on polling, this evaluation method is seldom an option (at least not after the fact in the case of elections). Pursuing social change and creating environmental awareness were evident in the waste sorting project that followed the Sankey diagrams task, which could be described as a generic template for projects aimed at fostering mathematical literacy. However, fostering aimed at actively pursuing social change within the perspective of mathematical literacy requires a change of norms and traditions in school mathematics.

It is not the tasks, nor the activities in themselves, that foster mathematical literacy; all depend on how the teacher implements the tasks and how the teacher interacts with the students in the classroom (Doerr & Ärlebäck, 2015). Without the proper preparation and actions taken by the teacher, any good and well-designed task can be spoiled and made into an uninteresting, boring routine exercise sitting comfortably in an established learning practice that focuses on procedural knowledge – classroom norms and ways of working that are counterproductive for fostering mathematical literacy. It is necessary that students experience how mathematics can be used to solve problems that interest them and are important to them in their social situation. Hence, teachers must work toward revised classroom norms in order to allow for more collaboration and consultation, creativity, students agency, etc. Within the waste sorting project, the teacher and the students in close interplay grasped the opportunity to highlight and explicate the use of mathematics in a social situation and in particular as a tool to convince the headteacher to buy more bins for improving their school environment. The students succeeded in achieving their goal largely due to the fostering instructional approach adapted by their teacher. By being encouraging and supportive as well as providing suggestions in line with what the students asked for, the teacher gave students enough time, resources, and agency to use mathematics in a productive and meaningful way. The balance between providing too much or too little instruction is of importance in the design and implementation of mathematical literacy tasks. The polling task was exploratory in nature with very little instructions given, and the pandemic task was highly structured with respect to instructions, whereas the instructions in the Sankey diagrams task tended to be more balanced and resulted in capturing and possibly fostering more of Jablonka's (2003) components.

Coming to the point, drawing on the tasks and analyses presented in this chapter, we see the following fostering principles as viable for the planning of activities that involve mathematical literacy that foster students' mathematical abilities to participate effectively and democratically in society in a productive and critical manner.

1. *Voice and choice*

   For fostering how individuals can improve their daily situations or solve problems in their social life, it is important to encourage students to develop agency by providing them with learning opportunities to discuss different options.

2. *Collaboration and time*

   For fostering team work so one can transform society in togetherness, it is important that the fosterer gives students ample time to collaborate in different forms with peers, teachers and even actors outside school.

3. *Engaging and meaningful contexts*

   For fostering students' inspiration to explore and learn about their social contexts, it is important that teachers design authentic tasks in which the contexts have intrinsic meaning for the students.

4. *Provoking and challenging*

   For fostering reflective and critical citizens, it is important that the fostering teacher asks questions that do not necessarily have a right or wrong answer as an eyeopener for different perspectives in the debate.

5. *Providing support and feedback*
For fostering students' road to becoming a contributing and participating citizen, it is important that the fostering teacher gives suggestions and feedback for how to proceed, which is a central part of the fostering process of mathematical literacy.

In this chapter we have shown that it is possible to foster mathematical literacy in connection with mathematical modelling. Based the three examples provided from our research on the teaching and learning of mathematical modelling, we argue that this connection will be more easily and effectively achieved if teachers and teacher educators adopt the role of fosterers. A first step in this direction might be to try and build the everyday mathematics classroom and teaching practices in line with principles like those described above.

References

Ärlebäck, J. B. (2009). On the use of realistic Fermi problems for introducing mathematical modelling in school. *The Montana Mathematics Enthusiast, 6*(3), 331–364. https://doi.org/10.54870/1551-3440.1157

Ärlebäck, J. B., & Frejd, P. (2021). Eliciting pre-service secondary teachers' initial ideas of sampling. In Y. Liljekvist, L. Boistrup, J. Häggström, L. Mattson, O. Olande, & H. Palmér (Eds.), *Hållbar matematikundervisning i en digitaliserad värld.: Proceedings of Madif 12, Växjö, 14–15, January, 2020* (pp. 234–244). SMDF.

Ärlebäck, J. B., & Frejd, P. (2020). The clash of cultural affordances: Using cultural affordances to analyze students' work in an activity using a mobile game app to simulate a pandemic outbreak. *Journal of Mathematics and Culture, 14*(2), 111–131.

Ärlebäck, J. B., Frejd, P., & Doerr, H. (2021). Exploring students' models of sampling and inference with nominal variables. *Quadrante: Revista de Investigação em Educação Matemática, 30*(1), 158–177. https://www.researchgate.net/publication/355107830_Exploring_students'_models_of_sampling_and_inference_with_nominal_variables

Blomhøj, M., & Hoff Kjeldsen, T. (2006). Teaching mathematical modelling through project work – Experiences from an in-service course for upper secondary teachers. *ZDM, 38*(2), 163–177.

Blomhøj, M., & Jensen, T. H. (2003). Developing mathematical modelling competence: Conceptual clarification and educational planning. *Teaching Mathematics and Its Applications, 22*(3), 123–139.

Boaler, J. (1998). Alternative approaches to teaching, learning and assessing mathematics. *Evaluation and Program Planning, 21*(2), 129–141. https://doi.org/10.1016/S0149-7189(98)00002-0

Bolstad, O. H. (2023). Lower secondary students' encounters with mathematical literacy. *Mathematics Education Research Journal, 35*, 237–253. https://doi.org/10.1007/s13394-021-00386-7

Borromeo Ferri, R. (2006). Theoretical and empirical differentiations of phases in the modelling process. *ZDM, 38*(2), 86–95. https://doi.org/10.1007/BF02655883

Brousseau, G. (2002). *Theory of didactical situations in mathematics: Didactique des Mathématiques, 1970–1990*. Springer.

Brown, M., Brown, P., & Bibby, T. (2008). "I would rather die": Reasons given by 16-year-olds for not continuing their study of mathematics. *Research in Mathematics Education, 10*(1), 3–18. https://doi.org/10.1080/14794800801915814

Cambridge Dictionary. (n.d.). Foster. Retrieved January 8, 2022, from https://dictionary.cambridge.org/dictionary/english/foster

Chapman, K. (2021). "Wait – it's a math problem, right?": Negotiating school frames in out-of-school places. *Educational Studies in Mathematics*. https://doi.org/10.1007/s10649-021-10099-0

Chih, T. (2021). Fostering collaboration in an asynchronous IBL modern geometry course. *Primus: Problems, Resources & Issues in Mathematics Undergraduate Studies*, 1–25. https://doi.org/10.1080/10511970.2021.1955781

D'Ambrosio, U. (1985). Ethnomathematics and its place in the history and pedagogy of mathematics. *For the Learning of Mathematics, 5*, 44–48.

Doerr, H. M., & Ärlebäck, J. B. (2015). Fostering students' independence in modelling activities. In K. Krainer & N. Vondrová (Eds.), *Proceedings of the 9th congress of European research in mathematics education* (pp. 855–862). Czech Republic, Prague.

Erbaş, A. K., Çetinkaya, B., Alacacı, C., Çakıroğlu, E., Aydoğan Yenmez, A., Şen Zeytun, A., Korkmaz, H., Kertil, M., Didiş, M. G., Baş, S., & Şahin, Z. (2016). *Lise matematik konuları için günlük hayattan modelleme soruları [Everyday life modeling questions for high school math subjects]*. TÜBA, Turkish Academy of Sciences.

Frejd, P., & Ärlebäck, J. B. (2017). Initial results from an intervention: affordances and constraints of using a mobile game app. In G. A. Stillman, W. Blum, & G. Kaiser (Eds.), *Crossing and researching boundaries in mathematical modelling education* (pp. 517–527). Springer. https://doi.org/10.1007/978-3-319-62968-1_43

Frejd, P., & Geiger, V. (2017). Exploring the notion of mathematical literacy in school curriculum documents. In G. A. Stillman, W. Blum, & G. Kaiser (Eds.), *Crossing and researching boundaries in mathematical modelling education* (pp. 255–263). Springer.

Frejd, P., & Vos, P. (2022). A commentary on the Special Issue "Innovations in measuring and fostering mathematical modelling competencies". *Educational Studies in Mathematics, 109*(2), 455–468. https://doi.org/10.1007/s10649-021-10122-4

Fyhn, A., & Robertsen, H. (2020). Kystfiskermatematikk og skolematematikk: To ulike perspektiver på hva ei méd er. *Nordic Studies in Mathematics Education, 25*(1), 65–88.

Gainsburg, J. (2008). Real-world connections in secondary mathematics teaching. *Journal of Mathematics Teacher Education, 11*(3), 199–219. https://doi.org/10.1007/s10857-007-9070-8

Geiger, V., & Frejd, P. (2015). A reflection on mathematical modelling and applications as a field of research: Theoretical orientation and diversity. In G. A. Stillman, W. Blum, & M. S. Biembengut (Eds.), *Mathematical modelling in education research and practice: Cultural, social and cognitive influences* (pp. 161–171). Springer.

Gould, H., Murray, D. R., & Sanfratello, A. (2012). *Mathematical modeling handbook*. COMAP.

Grootenboer, P., & Hemmings, B. (2007). Mathematics performance and the role played by affective and background factors. *Mathematics Education Research Journal, 19*(3), 3–20. https://doi.org/10.1007/BF03217459

Haara, F. O. (2018). Pedagogical entrepreneurship in school mathematics: An approach for students' development of mathematical literacy. *International Journal for Mathematics Teaching and Learning, 19*(2), 253–268.

Hernandez-Martinez, P., & Vos, P. (2018). "Why do I have to learn this?" – A study from mathematical modelling education about the relevance of mathematics. *ZDM Mathematics Education, 50*(1–2), 245–257. https://doi.org/10.1007/s11858-017-0904-2

Holenstein, M., Bruckmaier, G., & Grob, A. (2020). Transfer effects of mathematical literacy: An integrative longitudinal study. *European Journal of Psychology of Education, 1*–27. https://doi.org/10.1007/s10212-020-00491-4

Jablonka, E. (1996). *Meta-Analyse von Zugängen zur mathematischen Modellbildung und Konsequenzen für den Unterricht* [Doctoral dissertation]. Transparent Verlag.

Jablonka, E. (2003). Mathematical literacy. In A. Bishop, M. A. Clements, C. Keitel, J. Kilpatrick, & F. S. K. Leung (Eds.), *Second international handbook of mathematics education* (pp. 75–102). Kluwer Academic.

Kokkinos, J. (2009). *Does prior-to-school mathematics exist in practice?* [Paper presentation]. Third International Conference on Science and Mathematics Education (CoSEMEd), Penang, Malaysia. https://www.academia.edu/428099/DOES_PRIOR-TO-SCHOOL_MATHEMATICS_EXIST_IN_PRACTICE

Lave, J. (1988). *Cognition in practice: Mind, mathematics and culture in everyday life*. Cambridge University Press.

Lesh, R. A., & Doerr, H. M. (Eds.). (2003). *Beyond constructivism: Models and modeling perspectives on mathematics problem solving, learning, and teaching*. Lawrence Erlbaum Associates.

LEMA. (2009). *Reality-based tasks for school.* International Centre for STEM Education. https://www.matematikksenteret.no/sites/default/files/attachments/Booklet%20LEMA-prosjekt.pdf

Lerman, S. (2012). Culturally situated knowledge and the problem of transfer in the learning of mathematics. In L. Burton (Ed.), *Learning mathematics* (pp. 107–121). Routledge.

Maaß, K. (2008). *Mathematisches Modellieren – Aufgaben für die Sekundarstufe I.* Cornelsen Scriptor.

Munakata, M., & Vaidya, A. (2013). Fostering creativity through personalized education. *Primus, 23*(9), 764–775. https://doi.org/10.1080/10511970.2012.740770

Niss, M., & Blum, W. (2020). *The teaching and learning of mathematical modelling.* Routledge.

Perrenet, J., & Zwaneveld, B. (2012). The many faces of the mathematical modeling cycle. *Journal of Mathematical Modelling and Application, 1*(6), 3–21.

Prieto-Rodriguez, E. (2016). "It just takes so much time!" A study of teachers' use of ICT to convey relevance of mathematical content. *International Journal for Technology in Mathematics Education, 23*(1), 13–25. https://doi.org/10.1564/tme_V23.1.02

Sheldrake, R., Mujtaba, T., & Reiss, M. J. (2017). Science teaching and students' attitudes and aspirations: The importance of conveying the applications and relevance of science. *International Journal of Educational Research, 85*, 167–183. https://doi.org/10.1016/j.ijer.2017.08.002

Skolverket. (2012). *The subject syllabus for mathematics.* https://www.skolverket.se/download/18.4fc05a3f164131a74181063/1535372298267/Mathematics-swedish-school.pdf

Sokolowski, A. (2015). The effects of mathematical modelling on students' achievement-meta-analysis of research. *IAFOR Journal of Education, 3*(1), 93–114.

Verschaffel, L., Greer, B., & De Corte, E. (2000). *Making sense of word problems.* Lisse. Swets & Zeitlinger.

Vos, P. (2019). *Sankey diagrams.* University of Agder. https://www.academia.edu/41095813/Sankey_diagrams

Vos, P., & Frejd, P. (2022). The modelling cycle as analytic research tool and how it can be enriched beyond the cognitive dimension. In J. Hogdsen, E. Geraniou, G. Bolondi, & F. Ferretti (Eds.), *Proceedings of the 12th Conference on European Research in Mathematics Education (CERME-12), Bolzano Italy* (pp. 1185–1192). https://hal.archives-ouvertes.fr/hal-03759063

Vos, P., & Frejd, P. (2022). Grade 8 students appropriating Sankey diagrams: The first cycle in an educational design research. *Journal on Mathematics Education, 13*(2), 289–306. https://doi.org/10.22342/jme.v13i2.pp289-306

Vos, P., & Frejd, P. (2020). The object-tool duality in mathematical modelling – a framework to analyze students' appropriation of Sankey diagrams to model dynamic processes. *Avances de Investigación en Educación Matemática, 17*, 52–66. https://doi.org/10.35763/aiem.v0i17.305

Yackel, E., & Cobb, P. (1996). Sociomathematical norms, argumentation, and autonomy in mathematics. *Journal for Research in Mathematics Education, 27*(4), 458–477. https://doi.org/10.2307/749877

PART 4

Commentary

CHAPTER 9

Mathematical Literacy as a Disruptive Innovation

Merrilyn Goos

Abstract

In this chapter I reflect on some of the challenges of developing school students' mathematical literacy (ML) that have been touched on in the preceding chapters. To do this, I use an ML-flavoured model of the curriculum design and enactment process that distinguishes between the official curriculum, the operational curriculum, and mediating factors that influence elements of both. Along the way, I identify three ways in which mathematical literacy could be regarded as a disruptive innovation. (1) Mathematical literacy is inherently interdisciplinary and cannot be contained within the disciplinary confines of the official curriculum; (2) In the operational curriculum, teaching for mathematical literacy needs to foster students' critical thinking and gather evidence of a broader range of learning outcomes than are typically specified in the official curriculum; (3) The emphasis within initial teacher education on preparing discipline experts leaves too little room for developing cross-curricular ML knowledge. These observations are meant to encourage researchers and teachers to embrace the transformative possibilities of embedding mathematical literacy across the school curriculum.

Keywords

mathematical literacy – numeracy – curriculum design and enactment – official curriculum – operational curriculum – disruptive innovation

1 Introduction

This book offers a rich collection of approaches for developing school students' mathematical literacy (ML). The contributions come from many countries around the world, acknowledging the widespread influence of educational policies that aim to promote ML or assess students' ML competencies. In these chapters, teachers, teacher educators, and researchers will surely find ideas they can use or adapt for their own professional practice. My purpose in this commentary chapter, however, is to bring these contributions into some of the broader

 | DOI:10.1163/9789004710962_009

debates about mathematical literacy in school education. To do this, I will create a conceptual framework to help me identify issues in teaching ML and pose questions for deepening the ML research field. Because ML can be regarded as an innovation that disrupts traditional curriculum structures and pedagogical norms (cf. Falloon et al., 2021), my framework draws on Remillard and Heck's (2014) model of the curriculum policy, design, and enactment system. However, Remillard and Heck do not specifically discuss mathematical literacy and so I will also "flavour" my discussion framework with ML-related insights, drawing on the model of numeracy in the 21st century that I developed with colleagues to be used by school teachers across the curriculum (Goos et al., 2014).

2 Conceptual Framework

Remillard and Heck (2014) defined curriculum as "a *plan for the experiences* that learners will encounter, as well as the *actual experiences* they do encounter, that are designed to help them reach specified [mathematics] objectives" (p. 707, original emphasis). This definition indicates that a curriculum is more than a list of topics or learning objectives, and it highlights the well-known distinction between what the curriculum intends and what actually happens in classrooms. Remillard and Heck synthesised research on curriculum development to examine relationships between curricular intent and educational outcomes as well as how different actors reformulate curriculum at different levels within an educational system. Their resulting model (shown in Figure 9.1) depicts the features of a broader curriculum policy, design, and enactment system.

The *official* curriculum is specified by governing authorities and sets out expectations for students' learning. There are three components in the official curriculum: curricular aims and objectives, the content of consequential (or high stakes) assessments, and the designated curriculum. This latter component refers to the instructional plans, textbooks and other materials that, in some countries, might be endorsed or specified by a ministry of education to offer guidance towards addressing the curriculum's goals. As a designated curriculum does not exist in many countries, I will not consider this component of the official curriculum in any more detail.

The *operational* curriculum includes the teacher-intended curriculum, the enacted curriculum, and student outcomes. Thus, the operational curriculum represents the transformation of the official curriculum into teachers' personal plans, whether these are in writing or in the teachers' minds, and how these plans play out in the interactions between teachers and students in the classroom.

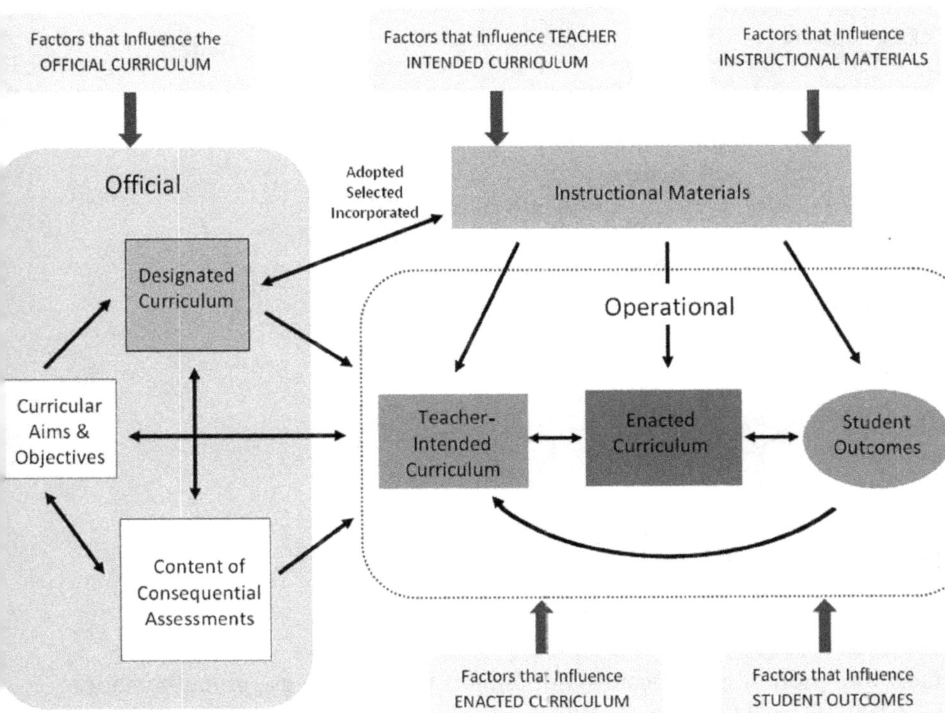

FIGURE 9.1 The curriculum policy, design, and enactment system (Remillard & Heck, 2014, p. 709)

Around the perimeter of their model, Remillard and Heck (2014) pointed to factors that influence elements of the official curriculum and the operational curriculum: these factors "may be social, political, cultural, structural, or cognitive" (p. 714). Influencing factors may include societal needs, values, expectations, and beliefs; views of individuals and groups wielding power; research on learning, teaching, and assessment; teacher knowledge, beliefs and practices; teachers' access to resources and support; contextual opportunities and constraints; and a range of student characteristics and cultural resources. These factors interact with each other in complex ways, and their degree of influence on the curriculum system may be either direct or subtle.

To understand some of the implications of incorporating mathematical literacy into curriculum design and enactment, I refer to the model of 21st century numeracy shown in Figure 9.2 (Goos et al., 2014). The model was developed to respond to educational policies in Australia that represent numeracy as a cross-curricular competency that should be developed by all teachers, no matter what subjects they teach. The numeracy model consists of four core elements: attention to real-life and curricular *contexts*; application of *mathematical knowledge*; use of physical, representational and digital

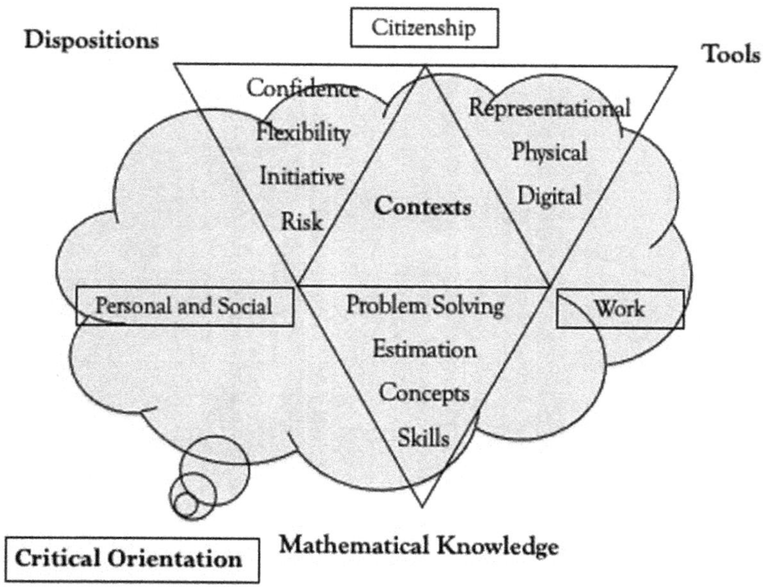

FIGURE 9.2 Numeracy model (Goos et al., 2014)

tools; and promotion of positive *dispositions* towards the use of mathematics to solve real-world problems. A fifth overarching element – a *critical orientation* – requires the appropriate selection and application of mathematics to a real-world problem as well as the interpretation and critique of results. The elements of the model are grounded in research on the uses of mathematical knowledge in workplace and other real-life contexts (e.g., Noss et al., 2000; Straesser, 2007); how various kinds of tools can represent mathematical concepts and support problem solving (e.g., Drijvers & Weigand, 2010; Hoyles et al., 2010); affective responses to mathematics (Attard et al., 2016); and interpretive, evaluative, and analytical aspects of being numerate (Ernest, 2002). Together, these elements portray a rich conceptualisation of numeracy (or mathematical literacy) in both real world and curricular contexts.

3 How Is Mathematical Literacy Represented in the Official Curriculum?

In the official curriculum, we would expect mathematical literacy to be represented in some way in the *curricular aims and objectives*. However, mathematical literacy is inherently interdisciplinary and embedded in an extra-mathematical context. The PISA mathematical literacy framework refers to formulating, using, and interpreting mathematics to solve real world problems, which inevitably transcend disciplinary boundaries. In schools, contexts for mathematical literacy are provided by other subjects in the curriculum such as science, art, and sport (see Chapter 5 by Leikin & Boriskovsky, this volume). Even when mathematical literacy is taught as a standalone subject in the school curriculum, as in South Africa (Pillay & Bansilal, Chapter 3 this volume), learners are meant to use mathematics to explore and solve problems in authentic, real-life contexts. In all these situations, the learning of mathematical literacy cannot be fully contained within the conventional boundaries of the school mathematics curriculum.

This observation highlights the tension between curriculum structures organised around disciplinary content (e.g., in subjects such as mathematics or science) and the interdisciplinary aspirations associated with cross-curricular competencies like mathematical literacy. In the former case, disciplinary aims and objectives are typically expressed at a level of detail that facilitates their transformation by teachers into classroom enactment. On the other hand, curricular aims for mathematical literacy are often expressed in more general terms that acknowledge the range of contexts in which ML can be developed: for example, the Australian Curriculum states that "students become numerate [i.e., mathematically literate] as they develop the knowledge and skills to use mathematics confidently across other learning areas at school and in their lives more broadly" (Australian Curriculum, Assessment and Reporting Authority [ACARA], n.d.). Statements such as this highlight the interplay between two dimensions of the 21st century numeracy model – *mathematical knowledge* and *contexts* – when formulating curricular aims for mathematical literacy.

In fact, *context* is the feature of mathematical literacy that distinguishes it from mathematics:

> Mathematics climbs the ladder of abstraction to see, from sufficient height, common patterns in seemingly different things. Abstraction is what gives mathematics its power; it is what enables methods derived from one context to be applied in others. But abstraction is not the focus of numeracy. Instead, numeracy clings to specifics, marshalling all relevant aspects of setting and context to reach conclusions. (Quantitative Literacy Design Team, 2001, pp. 17–18)

However, as the chapter by Almuna and Stacey (Chapter 4 this volume) shows, different levels of context use can increase the challenge of mathematical literacy tasks, and students do not always know how or whether to draw on their knowledge of context as a source of information for solving a problem.

Now let me turn to a more implicit way of representing mathematical literacy in the official curriculum, through the *content of consequential assessments*. Remillard and Heck (2014) argued that curriculum can be shaped by high-stakes assessments that are designed to track students' progress or measure the quality of an education system. While some such assessments might be implemented within individual countries, there can be no doubt about the much wider influence of the OECD's Programme for International Student Assessment (PISA). The PISA assessment programme and its definition of mathematical literacy were mentioned in almost every chapter of this book, and the "PISA shock" phenomenon led to curriculum reforms in countries such as Germany (OECD, n.d.), Norway (Bolstad & Goodchild, Chapter 1 this volume) and Ireland (O'Sullivan & Goos, Chapter 2 this volume) that served to raise the profile of mathematical literacy and its place in the school curriculum.

The issues discussed so far in this section should remind us that there is no easy way to represent mathematical literacy in the official curriculum's aims and objectives to convey its interdisciplinary connections with other subjects in the curriculum and real-world contexts beyond school. Nor can any country ignore the global influence of PISA's assessments of mathematical literacy on national curricula. This is why I refer to mathematical literacy as a *disruptive innovation*: because it challenges us to imagine new curricular structures and pedagogical approaches that will help students gain deep disciplinary knowledge as well as more integrative, cross-disciplinary capabilities.

4 How Is Mathematical Literacy Developed by the Operational Curriculum?

When teachers begin to transform descriptions of curricular objectives into personal teaching plans and classroom enactment, it is "the interactions between teachers and students around the tasks of each lesson and accumulated lessons in a unit of instruction" (Remillard & Heck, 2014, p. 711) that become important.

This book offers many examples of tasks and task design principles for developing students' mathematical literacy (e.g., see Chapter 4 by Almuna & Stacey, Chapter 5 by Leikin & Boriskovsky, Chapter 6 by Sawatzki et al., & Chapter 7 by Ärlebäck et al.). Many of these tasks involve using representational, physical, or digital *tools*, which aligns with the *tools* dimension of the numeracy model shown in Figure 9.2. There is perhaps less explicit focus on "instructional interactions and the norms that govern them", which Remillard and Heck (2014, p. 712) claim to be one of the most prominent dimensions of the enacted mathematics curriculum. However, this dimension is at the centre of the study reported by Frejd et al. (Chapter 8 this volume), who argue that traditional direct instruction, in which the teacher shows or tells students how to do a task, does not help students become mathematically literate. Instead, they propose a more implicit teaching approach that "fosters" mathematical literacy; that is, teachers activate students' own interests and provide them with opportunities for exploration, collaboration, and critical decision-making. Frejd et al. claim that mathematical modelling tasks provide a gateway to mathematical literacy, and yet only one of the modelling tasks that they used was deemed effective in achieving this goal. They attribute this success to a combination of task features, teacher actions, and classroom norms that fostered student agency. The effects of this mix of features are evident when students in their study not only succeeded in modelling a waste recycling problem but also used their model to persuade the school principal to obtain more recycling bins for the school yard. Frejd et al. argue that mathematical literacy can only flourish in classrooms governed by democratic norms that give students choices, time, challenge, and problems from daily life that they care about and want to solve.

Another component of the operational curriculum – student outcomes – refers to the products of the enacted curriculum (see Figure 9.1). In terms of the numeracy model presented in Figure 9.2, these outcomes include *mathematical knowledge*, positive *dispositions*, and a *critical orientation* towards using mathematics to solve real-world problems. But some of these outcomes are difficult to assess, especially if teachers must rely on written tests. Even in research projects designed to promote the rich interpretation of mathematical literacy depicted in Figure 9.2, teachers often struggle to find ways of evidencing, for example, their students' changing dispositions or inclination to think critically (Goos et al., 2019). One subtle aspect of mathematical literacy that is not captured by the model in 9.2 is its spontaneous nature – in real life, our mathematical literacy capabilities are typically called on at unpredictable times rather than required to be demonstrated "on demand", as in a test. The impromptu nature of mathematical literacy makes it inherently resistant to formal testing, but may instead open up possibilities for teachers to notice and document evidence of their students' learning in more informal ways (e.g., annotated work samples, photographs, video recordings).

Within the enacted curriculum, then, mathematical literacy can be a *disruptive innovation* that challenges teachers to foster students' critical thinking and to gather evidence of a broader range of learning outcomes than are typically specified in the official curriculum.

5 What Knowledge Do Teachers Need to Engage in Mathematical Literacy Practices across the Curriculum?

In the Introduction chapter to this book, Bolstad and Goodchild raised the question of how prospective teachers should be prepared to teach for mathematical literacy throughout their teacher education programme. They suggested that scant attention is being given to developing teachers' knowledge and competence in this area. In this book, the chapter by Pillay and Bansilal describes the lack of preparation received by teachers of the Mathematical Literacy subject in the South African school curriculum. In another chapter, O'Sullivan and Goos report that prospective teachers in Irish universities demonstrated limited understanding of numeracy and how to embed it within their specialist subject areas, even though this is mandated by the official curriculum. Remillard and Heck (2014) identified *teacher knowledge* as one of the key mediating factors that influence the curriculum system – particularly the operational curriculum. But it is challenging for teacher educators to prepare future teachers who are not only specialists in teaching one or more subjects in the curriculum but also knowledgeable about mathematical literacy demands and opportunities within their disciplinary specialisms.

A vast amount of research has raised awareness of the nature and importance of teacher knowledge. In mathematics education, the model of Ball et al. (2008) is best known for building on the seminal work of Shulman (1986), who developed theoretical ideas regarding the importance of subject matter knowledge, pedagogical content knowledge and curricular knowledge. The main contribution of Ball et al. is to distinguish between the different types of subject matter knowledge and pedagogical content knowledge required for teaching mathematics. They divide subject matter knowledge into three sub-domains:

- Common Content Knowledge (CCK): the mathematical knowledge that any educated person should possess;
- Specialised Content Knowledge (SCK): knowledge of mathematics that is specific and unique to teaching and not needed by other professions;
- Knowledge at the Mathematical Horizon (KMH): knowledge of how the mathematics being taught at a particular time connects with topics taught in the past and future.

Ball et al. divided pedagogical content knowledge into a further three sub-domains:

- Knowledge of Content and Students (KCS): teacher's knowledge of students' capabilities and learning preferences;

- Knowledge of Content and Teaching (KCT): knowledge of how to sequence mathematical topics and choose examples and representations;
- Knowledge of Content and Curriculum (KCC): knowing where a mathematical concept sits within the curriculum and which tools and resources can be used to address curricular goals.

But teaching mathematics is not the same as teaching for mathematical literacy, because ML – according to many curriculum policies – needs to be developed through students' engagement with all subjects in the school curriculum. To address the question of what knowledge teachers need to engage with mathematical literacy practices across the curriculum, O'Sullivan (2022) drew on the Ball et al. (2008) model of teacher knowledge and the numeracy model of Goos et al. (2014). She merged elements of these models to develop a three-component framework for teacher knowledge (see Figure 9.3), with Numeracy Knowledge acting as a bridge between the two pillars of Subject Specific Knowledge and Pedagogical Content Knowledge. Because of its visual representation, the model is labelled the "N" Framework.

O'Sullivan (2022) provided examples of subject-based tasks that could be used to enhance students' numeracy capabilities. One task, taken from Ireland's Junior Certificate Examination in Geography in 2018, asked students to interpret two population pyramids showing contrasting back-to-back stem-and-leaf plots for different countries. This section of the curriculum requires

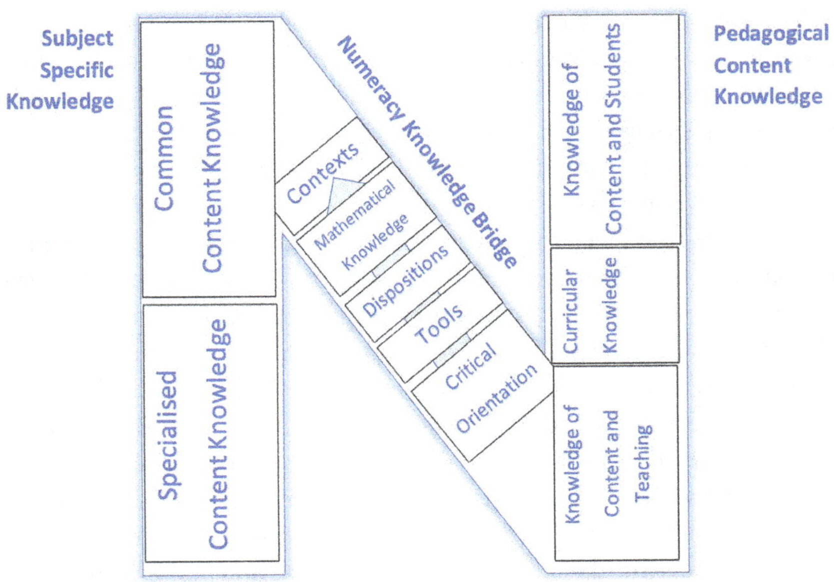

FIGURE 9.3 "N" Framework (O'Sullivan, 2022)

teachers to teach about topics such as globalisation, development, and population. Students apply their knowledge of these areas to explain settlement patterns, urbanisation, demographics, and human development (State Examination Commission, 2018). A population pyramid enables visualisation of the population distribution by age (e.g., in 5-year increments) and by sex (with the population frequency for males in each age group plotted on the left side of the pyramid and that for females on the right). (See National Geographic, 2023, for examples.)

This Geography task can be mapped to the "N" Framework to highlight the types of knowledge that a teacher would need to draw on to enhance students' understanding of the Geography topic and also their numeracy capabilities. With regard to Subject Specific Knowledge, a teacher would need Common Content Knowledge of which countries are identified as developed or developing, as well as Specialised Content Knowledge to enable comparison of life chances, health care, employment and education opportunities between developed and developing countries. In the Pedagogical Content Knowledge pillar of the "N" Framework, a teacher needs Curricular Knowledge of where population pyramids are situated in the Geography curriculum, Knowledge of Content and Students to recognise individual students' capabilities and whether they will be able to read population pyramids correctly. For Knowledge of Content and Teaching, a teacher must know how and why population pyramids are used to help students visualise patterns of population distribution. The Numeracy Bridge connects the two Knowledge Pillars, and should activate a teacher's knowledge of how to recognise numeracy demands and exploit numeracy opportunities in a curriculum-based task. In this example, the context is population distributions and mathematical knowledge of data representations is needed to make sense of the task. A teacher will have to know how to read this particular representational tool (back-to-back stem and leaf plots) and to feel confident of doing so (positive disposition). They will also need to exercise critical judgment to explain the contrasting shapes of the population pyramids and implications for social and economic conditions in the two countries they represent, and to guide students' attention to these features of the task.

O'Sullivan (2022) created the "N" Framework to classify the types of knowledge that teachers need to embed numeracy in subjects across the school curriculum. But she also realised the Framework could be used by teachers as an instrument for planning lessons and designing or modifying tasks to offer to students. There are likely to be other ways of representing knowledge domains relevant to numeracy/mathematical literacy teaching, but the "N" Framework may raise awareness of the complexity of embedding numeracy in curricular tasks. It could also be used by teacher educators to prepare future teachers in contexts where mathematical literacy is a cross-curricular competence. Because this appears to be a neglected area of initial teacher education, even in countries where all teachers are meant to develop their students' numeracy capabilities, such an approach would further support my claim that mathematical literacy is a *disruptive innovation* – not only in the realm of curriculum design and enactment, but also in teacher preparation.

6 Conclusion

Throughout this chapter I have described mathematical literacy as a potentially disruptive influence on curriculum, pedagogy, and teacher education. "Disruption" is not meant to convey negative connotations, but instead to highlight the transformative possibilities of embracing mathematical literacy as a guiding principle in educating young people for an uncertain world. Disruption can create opportunity to do things in new ways and break down barriers that hold people back from learning. For this reason, let me pose some questions for future research into mathematical literacy, arising from the ideas discussed in this chapter.

- What is (or should be) the rationale for including mathematical literacy in the school curriculum? How should it be represented so that teachers can see ways of exploiting the inherent mathematical literacy demands of the subjects they teach?
- What mathematical literacy learning outcomes should be valued? How should evidence of this learning be collected, analysed, and communicated?
- How should prospective teachers be prepared for developing their students' mathematical literacy, without shifting emphasis away from disciplinary learning? Which teacher educators should be responsible for developing these professional capabilities in teachers?

No doubt there are many more questions that could (and should) be asked. I leave it to the readers of this book to take up the disruptive challenges I have outlined here and to search for answers.

References

Attard, C., Ingram, N., Forgasz, H., Leder, G., & Grootenboer, P. (2016). Mathematics education and the affective domain. In K. Makar, S. Dole, J. Visnovska, M. Goos, A. Bennison, & K. Fry (Eds.), *Research in mathematics education in Australasia 2012–2015* (pp. 73–96). Springer. https://doi.org/10.1007/978-981-10-1419-2

Australian Curriculum, Assessment and Reporting Authority (ACARA). (n.d.). *Numeracy*. Retrieved November 6, 2023, from https://www.australiancurriculum.edu.au/f-10-curriculum/general-capabilities/numeracy/

Ball, D. L., Thames, M. H., & Phelps, G. (2008). Content knowledge for teaching: What makes it special. *Journal of Teacher Education, 59*(5), 389–407. https://doi.org/10.1177/0022487108324554

Drijvers, P., & Weigand, H. (2010). The role of handheld technology in the mathematics classroom. *ZDM Mathematics Education, 42*(7), 665–666. https://doi.org/10.1007/s11858-010-0285-2

Ernest, P. (2002). Empowerment in mathematics education. *Philosophy of Mathematics Journal, 15*(2). https://education.exeter.ac.uk/research/centres/stem/publications/pmej/pome15/empowerment.htm

Falloon, G., Stevenson, M., Beswick, K., Fraser, S., & Geiger, V. (2021). Building STEM in schools: An Australian cross-case analysis. *Educational Technology & Society, 24*(4), 110–122. https://www.jstor.org/stable/48629249

Goos, M., Geiger, V., & Dole, S. (2014). Transforming professional practice in numeracy teaching. In Y. Li, E. Silver, & S. Li (Eds.), *Transforming mathematics instruction: Multiple approaches and practices* (pp. 81–102). Springer. https://doi.org/10.1007/978-3-319-04993-9_6

Goos, M., Geiger, V., Dole, S., Forgasz, H., & Bennison, A. (2019). *Numeracy across the curriculum: Research-based strategies for enhancing teaching and learning*. Routledge. https://doi.org/10.4324/9781003116585

Hoyles, C., Noss, R., Kent, P., & Bakker, A. (2010). *Improving mathematics at work: The need for techno-mathematical literacies*. Routledge.

National Geographic. (2023). *Population pyramid*. Retrieved November 19, 2023, from https://education.nationalgeographic.org/resource/population-pyramid/

Noss, R., Hoyles, C., & Pozzi, S. (2000). Working knowledge: Mathematics in use. In A. Bessot & J. Ridgeway (Eds.), *Education for mathematics in the workplace* (pp. 17–35). Kluwer. https://doi.org/10.1007/0-306-47226-0

Organisation for Economic Cooperation and Development (OECD). (n.d.). *Germany's PISA shock*. Retrieved November 6, 2023, from https://www.oecd.org/about/impact/germany-pisa-shock.htm

O'Sullivan, K. O. (2022). *Investigating pre-service teachers' knowledge of numeracy and their ability to teach numeracy for disciplinary learning*. [Doctoral dissertation, University of Limerick]. University of Limerick Research Repository. https://doi.org/10.34961/researchrepository-ul.21591333.v1

Quantitative Literacy Design Team. (2001). The case for quantitative literacy. In L. Steen (Ed.), *Mathematics and democracy: The case for quantitative literacy* (pp. 1–22). National Council on Education and the Disciplines.

Remillard, J., & Heck, D. (2014). Conceptualizing the curriculum enactment process in mathematics education. *ZDM Mathematics Education*, *46*, 705–718. https://doi.org/10.1007/s11858-014-0600-4

Shulman, L. S. (1986). Those who understand: Knowledge growth in teaching. *Educational Researcher*, *15*(2), 4–14. https://doi.org/10.3102/0013189X015002004

State Examination Commission. (2018). *Exam material archives*. Retrieved November 19, 2023, from https://www.examinations.ie/exammaterialarchive/?i=110.112.93.101.96.94.111.103

Straesser, R. (2007). Didactics of mathematics: More than mathematics and school! *ZDM Mathematics Education*, *39*(1), 165–171. https://doi.org/10.1007/s11858-006-0016-x

Index

agency 2, 39, 72, 140, 142, 185, 197, 199, 213
analysis
  narrative 52
  thematic 34, 35
arithmetic 4, 6, 7, 48
authentic/authenticity 2, 14, 16, 50, 63, 116, 117, 143, 144, 172, 192, 196, 199, 211

boundary 7, 49
  crossing 49

capability 31, 128, 129
  financial 15, 16, 126, 128, 129, 132, 133, 137, 141, 142, 145, 146
common European numeracy framework 48
competence/competencies
  cross curricular 3, 24, 28, 209, 211, 217
  mathematical 3, 4, 15, 86, 89, 90, 97, 119, 154, 180
  teaching 3, 8, 9, 11
computation 6, 155
conceptualization 3, 7, 27, 74, 86, 155, 210
content knowledge
  curricular 11, 131
  mathematical 11, 145
  pedagogical 13, 16, 131, 133, 138, 145, 214–216
  specialised 132, 214, 216
context/contextual
  concepts 51
  language 49, 51, 58, 64, 65
  real–life 27, 49, 50, 55, 209–211
  real world 3, 14, 15, 24, 27, 36, 41, 72, 74, 76, 78, 85–91, 98, 130, 212
  rules 49–51, 56, 61, 62, 64, 65
COVID-19 4, 77, 132, 146, 183, 190, 192, 193
curriculum/curricula
  design 17, 209, 217
  documents 6, 10, 12, 47, 50
  enactment 17, 208, 209, 217
  integration 28
  national 3, 9, 212
  Norwegian 5, 9
  official 26, 208–210, 212–214
  operational 208, 209, 212–214
  requirements 11, 13, 15, 30, 41, 132
  teacher education 8, 9, 12, 30, 41

design
  research 16, 32, 126, 131, 132
  tasks 41, 95, 134, 212
development
  practices 2, 10, 11, 126, 128, 144, 154, 184
  professional 8, 11–13, 15–17, 26, 62–65, 115, 154
  students' mathematical literacy 16, 50, 52, 62, 100, 101, 125, 166, 180, 184, 207, 212
  understanding 2, 8, 10, 11, 25, 42, 62, 126, 195
disposition 16, 27, 48, 98, 154, 156, 158–161, 165, 167, 169, 172, 180, 210, 213
domain
  conceptual 50
  contextual 50, 51

education
  citizenship 4, 6, 154
  teacher 1, 2, 5, 8, 9, 11–13, 17, 25, 28, 30–34, 38, 39, 41, 42, 130, 131, 179, 186, 214, 217
  for mathematical literacy 2, 4, 13, 31, 39, 42, 72, 74, 77, 127, 132, 153, 159 179, 214
  for teaching mathematical literacy 50, 216
  in-service 2
  pre-service/initial 12, 13
enactment 17, 135, 208, 209, 211, 212, 217
experience(s) 1, 2, 6, 8, 10–12, 14, 15, 24, 25, 30, 34, 37, 39–42, 65, 72, 88, 89, 95, 116, 118, 130, 133, 144–146, 157, 164, 166, 169, 170, 184, 188, 198, 199, 208
  practice–based 2

foster 5, 28, 180, 185, 186, 189, 190, 192, 193, 198–200, 213

geometry 4, 82, 114, 115, 118
graphs 4, 29, 34, 36, 37, 53–55, 75, 80, 89, 104, 106, 110, 111, 130, 162, 183, 191–193
guidance 3, 10, 13, 25, 26, 40, 129, 131, 145, 208

implement 5, 9, 10–13, 18, 25, 28, 31, 116, 118, 153, 167, 172, 184, 186
  tasks 2, 3, 31, 41, 97, 100, 102, 114–116, 185, 187, 189, 190, 192, 193, 196, 198, 199, 212
innovation 17, 135, 207, 208, 212, 213, 217
inquiry 1
  interdisciplinary 28
intervention 1, 126, 133
investigation 28, 81, 91, 134, 135, 142, 156, 157, 166, 167, 171, 172

knowledge
  common 75, 76, 101
  curricular 131, 132, 138, 143, 214, 216
  disciplinary 26, 126, 212
  mathematical 4, 5, 15, 27, 34, 36–38, 41, 48, 72–74, 88, 100, 101, 128–130, 133, 142, 145, 154, 156, 159, 161, 165, 171, 180, 209–211, 213, 214, 216
  pedagogical content 13, 16, 131, 138, 145, 214–216
  procedural 48, 72, 198
  specialised 132, 214, 216

language 2, 13, 29, 33–35, 37, 38, 40, 49, 51, 57, 58, 64, 65, 73, 129, 135, 145, 162, 179, 183, 185, 187, 192
  complex 49, 50
learner 9, 31, 47–58, 60–64, 163, 185, 208, 211
  centered 63
learning 1, 2, 5–12, 14–18, 25–28, 31, 33, 39, 40, 42, 47, 48, 50, 62, 65, 77, 91, 95, 101, 115–116, 128–136, 139–146, 153, 155, 163–167, 172, 179, 184, 185, 187, 190, 191, 197–200, 208, 209, 211, 213, 214, 217
  opportunities 2, 6, 8–10, 39, 143, 146, 163, 165, 172, 199
literacy
  conceptual 15, 96, 97, 118
  extra-mathematical 15, 210
  contextual 14, 15, 17, 96, 97, 99, 100, 116, 118, 119
  mathematical 1–3, 7–9, 12–17, 23–26, 28, 30–32, 41, 46–48, 50, 52, 62, 72, 73, 81, 89, 90, 92, 95–102, 106, 110, 111, 115, 116, 118, 119, 125–127, 152, 153, 165, 166, 178–186, 189, 190, 192, 193, 197–200, 207–217
  quantitative 25, 30, 31, 48, 98, 159, 173
  statistical 16, 17, 48, 98, 153, 155, 159–161, 163, 165, 167, 173

mathematical
  communication 4
  competence/competencies 3, 4, 15, 73, 86, 89, 90, 97, 119, 154, 180
  knowledge 4, 5, 15, 27, 34, 36–38, 41, 48, 72–74, 88, 100–102, 128–130, 133, 142, 145, 154, 156, 159, 161, 165, 171, 180, 209–211, 213, 214, 216
  modelling 4, 14, 16, 17, 72, 74, 91, 99, 101, 116, 128, 133, 134, 141, 163, 178–185, 193, 196–198, 200, 213
  proficiency 15, 98
  reasoning 4, 49, 73, 127, 159, 181
  representation 4, 85, 197
  symbols 4
measuring 4, 103, 106, 155, 168
methods 4, 9, 28, 30–32, 51, 76, 81, 87, 132, 153, 157, 159, 161, 163, 172, 190, 198, 211
mathematics 1–10, 12–16, 24–35, 37–40, 42, 47–52, 54, 55, 57, 59, 62–65, 72–78, 81, 82, 85, 88–91, 95, 96, 98–102, 115–116, 126–130, 132, 134, 135, 137, 139–145, 153–155, 159, 163, 165–167, 170, 172, 173, 178–185, 187, 189–193, 195, 197–200, 210, 211, 213–215
  education 2, 4, 13, 31, 39, 42, 72, 74, 77, 127, 132, 153, 159, 179, 214
modelling 4, 14, 16, 17, 72, 74, 91, 99, 101, 116, 128, 133, 134, 141, 163, 178–185, 193, 196–198, 200, 213

numeracy 6, 7, 9, 13, 15, 17, 23–42, 48, 50, 63, 97, 98, 126–130, 132, 133, 135, 137, 139–142, 145, 154, 173, 208–217

opportunity to learn 6, 10, 30, 37, 39, 41
Organization for Economic Cooperation and Development/OECD 3, 14, 34, 46, 73, 78, 80, 81, 86, 91, 92, 98, 126, 127, 212
orientation 26–28, 101, 128, 131, 154, 158, 161, 165, 210, 213

PISA/Programme for International Student Assessment 3, 5, 13, 14, 17, 24, 26, 30, 72–77, 80–83, 85–87, 89–92, 98, 126, 127, 210, 212
policy 3, 5, 11, 13, 24, 30, 65, 126, 128, 133, 208, 209
problem solving 4, 24, 34, 35, 36, 47, 75, 81, 89–91, 96, 99, 100, 102, 118, 128, 134, 153, 156, 158, 159, 168, 180, 181, 183, 210

research
action 13, 26, 28
document analysis 52
focus group 34, 35, 39
interviews 8, 16, 32, 34, 35, 37, 39, 40, 52, 88, 137, 138
mixed methods 32, 81
observation 26, 42, 49, 52, 57, 89, 91, 197, 211
online surveys 31
qualitative 51
questionnaire 33–37, 39, 115
sampling 52, 156, 187–190
stimulated recall 52
reason 5, 13, 14, 61, 76, 98, 128, 159, 193, 195, 217
mathematically 3, 24, 47, 64, 98
reasoning 73, 90, 119, 127, 153, 155–157, 169–172
mathematical 4, 48, 49, 60, 73, 98, 111, 127, 159, 161, 162, 181, 189, 190, 193
resource(s)
digital 10
educational 9, 72

skills
basic 5, 6, 9, 47, 156
cross-curricular 5
key 26
standards
numeracy 31, 32
professional 30, 31
statistics 16, 36, 37, 87, 152, 153, 155, 156, 157, 159–161, 166, 170–173, 189, 190, 192
subject
area 7, 11, 12, 31, 32, 42, 131, 132, 214
carrier 28
specialism 25, 31–34, 38, 40–42, 214

task(s) 2–4, 8–12, 14–17, 29, 31, 33, 36, 41, 48–50, 52–55, 58–65, 88, 95, 97, 99–102, 106, 115, 116, 118, 119, 127–135, 140, 141, 144–146, 156, 167–169, 178–180, 184–199, 211–213, 215, 216
teachers/practitioners
beliefs 62, 65, 209
conceptions 62, 65, 97, 100, 114, 131
experienced 2, 39, 88, 170
knowledge 132, 139, 145, 214
practicing 1, 2
pre-service 1, 2, 12, 13, 23–25, 28, 30–41, 48, 186
specialist 14, 62, 63, 214
TEDS-M/Teacher Education and Development Study 30, 41
textbooks 6, 10, 11, 59–61, 64, 65, 134, 135, 143, 145, 184, 208
digital 10
TIMSS/Trends in International Mathematics and Science Study 5, 26, 30
tools
digital 27, 191–193, 212
physical 27, 180, 209, 212
representational 27, 180, 209, 212, 216
training 5, 9, 47, 52
in-service 47

vocation/vocational 4